Wolfgang Gentzsch

Vectorization of Computer Programs with Applications to Computational Fluid Dynamics

Notes on Numerical Fluid Mechanics
Volume 8

Series Editors: Ernst Heinrich Hirschel, München
Maurizio Pandolfi, Torino
Arthur Rizzi, Stockholm
Bernard Roux, Marseille

Volume 1 Boundary Algorithms for Multidimensional Inviscid Hyperbolic Flows (Karl Förster, Ed.)

Volume 2 Proceedings of the Third GAMM-Conference on Numerical Methods in Fluid Mechanics (Ernst Heinrich Hirschel, Ed.) (out of print)

Volume 3 Numerical Methods for the Computation of Inviscid Transonic Flows with Shock Waves (Arthur Rizzi / Henri Viviand, Eds.)

Volume 4 Shear Flow in Surface-Oriented Coordinates (Ernst Heinrich Hirschel / Wilhelm Kordulla)

Volume 5 Proceedings of the Fourth GAMM-Conference on Numerical Methods in Fluid Mechanics (Henri Viviand, Ed.) (out of print)

Volume 6 Numerical Methods in Laminar Flame Propagation (Norbert Peters / Jürgen Warnatz, Eds.)

Volume 7 Proceedings of the Fifth GAMM-Conference on Numerical Methods in Fluid Mechanics (Maurizio Pandolfi / Renzo Piva, Eds.)

Volume 8 Vectorization of Computer Programs with Applications to Computational Fluid Dynamics (Wolfgang Gentzsch)

Volume 9 Analysis of Laminar Flow over a Backward Facing Step (Ken Morgan / Jaques Periaux / François Thomasset, Eds.)

Volume 10 Efficient Solutions of Elliptic Systems (Wolfgang Hackbusch, Ed.)

Volume 11 Advances in Multi-Grid Methods (Dietrich Braess / Wolfgang Hackbusch / Ulrich Trottenberg, Eds.)

Volume 12 The Efficient Use of Vector Computers with Emphasis on Computational Fluid Dynamics (Willi Schönauer / Wolfgang Gentzsch, Eds.)

Volume 13 Proceedings of the Sixth GAMM-Conference on Numerical Methods in Fluid Mechanics (Dietrich Rues / Wilhelm Kordulla, Eds.)

Manuscripts should have well over 100 pages. As they will be reproduced photomechanically they should be typed with utmost care on special stationary which will be supplied on request. In print, the size will be reduced linearly to approximately 75%. Figures and diagrams should be lettered accordingly so as to produce letters not smaller than 2 mm in print. The same is valid for handwritten formulae. Manuscripts (in English) or proposals should be sent to the general editor Prof. Dr. E. H. Hirschel, MBB-LKE 122, Postfach 80 11 60, D-8000 München 80.

Wolfgang Gentzsch

Vectorization of Computer Programs with Applications to Computational Fluid Dynamics

With Contributions of

Roland Kessler
Wilhelm Kordulla
Eckart Meiburg
Bernhard Müller
Dieter Schwamborn

Friedr. Vieweg & Sohn Braunschweig/Wiesbaden

CIP-Kurztitelaufnahme der Deutschen Bibliothek

Gentzsch, Wolfgang:
Vectorization of computer programs with applications
to computational fluid dynamics/Wolfgang Gentzsch.
With contributions of Roland Kessler ... –
Braunschweig; Wiesbaden: Vieweg, 1984.
(Notes on numerical fluid mechanics; Vol. 8)
ISBN-13: 978-3-528-08082-2 e-ISBN-13: 978-3-322-87861-8
DOI: 10.1007/978-3-322-87861-8

NE: GT

Produced by Industrie u. Verlagsdruck, Walluf b. Wiesbaden

IODN-10: 078 3 528 08082 2

Preface

The scope of the present book is to offer the most efficient tools for the vectorization of serial computer programs. Here, by vectorization we understand the adaptation of computer programs to the special architecture of modern available vector computers to exploit fully their potential, which will often result in remarkable performance improvements. The book is written primarily for users working in the various fields of computational physics, for scientists as well as for programmers running their jobs on a vector computer. The text may, however, also be of value to those who are interested in numerical algorithms. Although the examples discussed in chapter 9 have been taken from Computational Fluid Dynamics, the numerical methods are well-known, and are applied in many fields of Computational Physics.

The book is divided into four parts. After a short introduction which outlines the limits of conventional serial computers in contrast to the possibilities offered by the new vector machines, the second part is addressed to the discussion of some main features of existing computer architectures. We restrict ourselves to the vector computers CRAY-1S and CDC-CYBER 205, although, in the meantime, many vector and parallel computers and array processors are available such as DENELCOR's Heterogeneous Element Processor (HEP), ICL's Distributed Array Processor (DAP), SPERRY UNIVAC's Array Processing System (APS), STAR TECHNOLOGIES ST-100, FLOATING POINT SYSTEMS' Array Processor (FPS), FUJITSU's FACOM VP-100 and VP-200, HITACHI's Integrated Array Processor (IAP), HITACHI's S 810/10 and S 810/20 and others. Many of the ideas developed in this book are transferable to these computers, too.

In the future, vectorization of simple DO-loops will be done by the auto-vectorizer. However, the right choice of suitable algorithms which fit the computer architecture, will always be important. Therefore, chapters 3 to 8 deal with vectorization of basic numerical algorithms. For that purpose an introduction into the principles of vectorization is given. After a discussion of different programming languages, the handling of loops and algorithms and the restructuring of important numerical methods such as Gaussian elimination and iterative methods is explained in detail.

Finally in chapter 9, the application of the previously developed ideas to real production codes for the solution of problems in Computational Fluid Dynamics is presented. The numerical methods of MacCormack, Beam and Warming and Richtmyer, a Galerkin procedure and a Monte-Carlo simulation have been applied to the basic differential equations for the solution of simple but realistic flow problems. The physical phenomena, the numerical methods and the problem of vectorization are explained in detail, and some results for serial and vector computers are given.

I wish to thank Prof. E.H.Hirschel who suggested to write this book, and the other editors of this series for their valuable discussions and suggestions concerning the manuscript. The book was prepared for publication with the cooperation of members of the staff of the Institute for Theoretical Fluid Mechanics of the DFVLR in Goettingen, see the contributions in chapter 9. I especially wish to acknowledge the interest and encouragement of the director of the Institute, Dr. H. Oertel. Assistance in calculation has been given by Klaus Flint and Bernd Will. Mike Dowling has helped greatly by reading the manuscript and suggesting important improvements. The typing has been most effectively handled by Karin Fichna, the formulae by Dietlind Schlesinger and the preparation of the figures by Evelyn Kraft. To all these individuals, and several others unnamed, my thanks. Finally, I wish to thank my wife, Manina, for her patient support during the writing of the manuscript.

November 1983

DFVLR, Bunsenstr. 10
D-3400 Goettingen, FRG

Wolfgang Gentzsch

Contents

Page

1. Introduction: Supercomputers in Computational Fluid Dynamics

Since the installation of the electronic computers ENIAC in Pennsylvania in 1946 and G1 in Goettingen 1950, the importance of computers for engineering and the natural sciences has grown continuously. In the same period, owing to improvements in technology and architecture, the performance rate of serial von Neumann computers has increased to several Million Floating Point Operations per Second (MFLOPS). However, since 1969 with the CDC-7600, no substantial increase in performance has been achieved for serial computers with conventional von Neumann architecture. The VLSI-technology (Very Large Scale Integrated circuits with more than 260000 transistors on one chip) is nearly fully developed. Future technologies such as GAAS (Gallium-Arsenid) or the super-conductive Josephson-Junction will cause an improvement in speed by a factor of about 10 to 20 but are not yet at our disposal.

The other way to significantly increase the performance rate of a computer is by improving the architecture. This led us to the so-called vector and parallel computers. In the early 1970's the ILLIAC IV was built by the Burroughs Corporation (see |7|,|88|,|121|, |130| and |152|). This parallel computer was over one order of magnitude faster parallel than conventional computers at that time. Modern vector computers such as the CRAY-1 and the CDC-CYBER 205 offer greatly enhanced computing speeds arising from the ability to operate in a pipelining fashion, while parallel computers perform many calculations simultaneously. The architectural features of these machines tend to favour certain classes of algorithms. Beyond that, a certain arrangement of data stored in memory and a certain structure of the computer code is necessary to avoid memory bank conflicts, and to guarentee a continuous stream of data from the memory to the CPU (Central Processing Unit). Taking these characteristics into account, the user will be able to exploit fully the potential of vector machines, which will often result in remarkable performance improvements.

Normally restructuring of existing codes is confined to only a few subroutines, namely the most time-consuming parts of a program. In scientific programs, for example, the most common problem is to solve a discrete analogue of the basic system of differential equations. Here, by far most of the computation time is spent in calculating the coefficients of the resulting algebraic system of equations, and in solving this system for each time or iteration level. Vectorization of these parts is often sufficient to take advantage of the high efficiency of the vector machines.

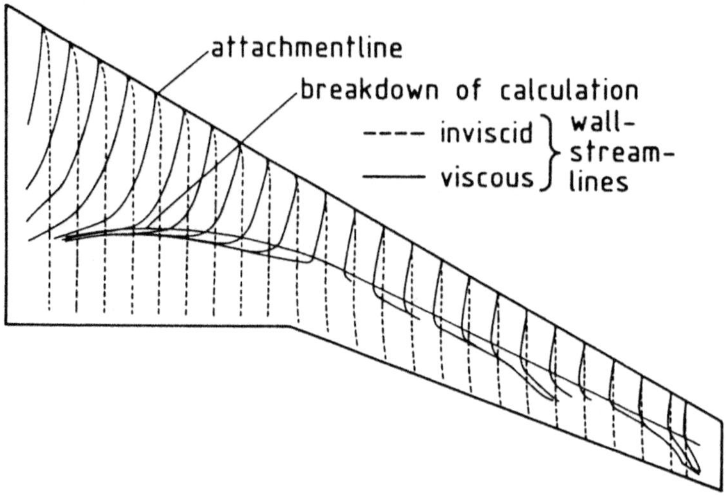

Fig. 1.1: Wall-Streamlines of viscous and inviscid flow on the lower surface of a wing |241|.

Using three examples we now outline the possibilities and limitations of numerical calculations in computational fluid dynamics using modern supercomputers.

The first example deals with the computation of a three-dimensional boundary-layer flow. The numerical solution of boundary-layer problems progressed considerably over the last 20 years (see e.g. |225|). One contribution to this relatively classical field is the paper of Schwamborn |241| (see also section 9.3) concerning the laminar boundary-layer on wings and wing-like bodies for the sake of hydrodynamic stability considerations (Fig. 1.1). In a realistic boundary-layer flow calculation including the attachment line, one has a computational grid of 120x40x50 points. Considering four iterations per grid point, we have about

120x40x4 = 19200 systems of linear equations in
5x50 = 250 unknowns for each grid line x_2= const.

The total storage requirement is about 1.65 Mio words if all the data are to be kept for later analysis. Computation time is about 1.2×10^{-3} sec per gridpoint on an IBM-3081, which results in 288 sec for the 120x40x50 grid and a compressible version of the code.

However, coupling the turbulent boundary-layer calculations with inviscid full -potential codes yielding numerical solutions within a certain accuracy, results in massive computation time and memory requirements. A three-dimensional calculation of an inviscid flow field past a wing by the transonic wing code FLO22 (|192|,|229|) on a relatively coarse grid of 192x32x16 points (chord x span x normal) uses about 2000 sec computer time on an IBM-3081. The inclusion of a boundary-layer calculation would increase this time to almost one hour. According to the coarse grid, some important properties of the flow field such as swept oblique shock waves are numerically reproduced incorrectly, but a calculation on a finer grid of e.g. 300x32x32 points far exceeds the capacity of modern serial computers.

It is only the successors of the serial supercomputers, the so-called vector computers, that are able of handling problems of this magnitude in a realistic time. To illustrate this we estimate the computational time and memory requirements for a Navier-Stokes solution of the supersonic flow field past the above mentioned blunt body (see section 9.2). A thin-layer approximation of the Navier-Stokes equations is solved by the difference method of Beam and Warming |217| on a 40x75x60 grid. The storage requirement per grid point would then be roughly |218|:

5 words for	ρ , ρu , ρv , ρw , e
5 words for	$\Delta\rho$, $\Delta\rho u$, $\Delta\rho v$, $\Delta\rho w$, Δe
10 words for	ξ_x, ξ_y, ξ_z, η_x, η_y, η_z, ζ_x, ζ_y, ζ_z, J.
20 words	

ρ here means density, u,v,w the velocity components in x-, y- and z-directions respectively, e energy per unit volume, $\Delta\rho$, ... the difference of the density for two neighbouring time levels, and so on, ξ_x ,..... , ζ_z the derivatives of the transformation, and J the Jacobi determinant. Then for the whole grid we would have a storage requirement of

40x75x60x20 = 3.6 Mio words

For this example, and the selected grid, the numerical computation takes about |218|:

1.2×10^{-3} sec per time step, per grid point,
3.6 min per time step and all grid points,
60 h for 1000 time steps and all grid points,

using an IBM-3081. On the vector computers CRAY-1 and CYBER 205 for example, the same calculations (1000 time steps, same grid) with a vectorized code would require only about 3 hours. A comparison of computation times for a similar problem on different computers is shown in Table 1.1 (for a detailed discussion see section 9.2). However, considering the low accuracy due to the coarse grid the times given in Table 1.1 are not very satisfactory for the computational fluid dynamicist.

Table 1.1: Comparison of computation times for different computers (DFVLR-benchmark |68,140|, CPU= Central Processing Unit).

Computer	CPU-time	
UNIVAC 1106	100	days
ICL 2980	25	"
CDC CYBER 175	4	"
IBM 3081 K	3	"
CDC 7600	2	"
CDC STAR 100	1	"
ICL DAP	13	hours
HITACHI S9/IAP	10	"
FUJITSU 7890	10	"
ILLIAC IV	6	"
CRAY - 1S	4	"
CYBER 205	4	"
CRAY - XMP	3	"

Successors to the present vector computers are now almost ready for their market release. In 1984 FUJITSU's FACOM VP 100/200 and HITACHI's S 810/10 and S 810/20 will go into production. They are all register-to-register machines, comparable in vector speed and architecture to the CRAY X-MP. In about 1985 the CRAY and CYBER successors will be expected both with peak MFLOPS rates (Millions of FLoating point OPerations per Second) of more than one thousand (= one GIGA-FLOPS).

At least for the next ten years the principle of vector processing through segmented functional units (called pipes) and over-lapped solution of more complicated vector operations will probably predominate. It is therefore necessary to devote time and effort to the problem of efficiently adapting scientific programs to the new vector architectures. For various fluid dynamic codes, the implementation on different vector computers has been discussed in references |15|, |19|, |35|, |65|, |68|, |79|, |110|, |129|, |130|, |133|, |140|, |152|, |156|, |159|, |176|, |184|, |185|, |186|, |192|, |193|, |194| and |195|. Moreover, a general discussion of vector and parallel algorithms may be found in |53|, |54|, |55|, |56|, |57|, |66|, |82|, |85|, |89|, |90|, |91|, |92|, |116|, |122|, |124|, |158|, |162|, |171|, |172|, |174|, |188|, |196|, |205|, |209|, |210| and |211|.

2. Computer Architectures

Enormous progress in computer performance is not only due to the immense improvement in semi-conductor and circuit technology, and in signal transfer technology, but also to improvements in computer architectures; a fact very important to the user, who must adapt his program in language and algorithm accordingly.

We devote this chapter therefore to a cursory survey of the main classes of computer architectures in use today (see e.g. |14|, |29|, |53|, |60|, |66|, |72|, |77|, |78|, |85|, |86|, |90|, |103|, |104|, |118|, |120|, |122|, |123|, |144|, |153|, |157| and |180|), namely

fundamental systems
multi-function systems
pipeline processors
parallel processors.

We will consider the following example (see |65|) from a user's perspective:

$$D = (A + B) * C ,$$

where A,B,C and D are vectors of length 64.

Deeper knowledge of the important architectural features is absolutely necessary for the vectorization of production codes. We therefore go into a more detailed discussion of the architecture of the two vector machines CRAY-1 and CDC-CYBER 205. These two computers have been selected for special attention, since:

- Together they represent the two principal architectures presently in use, the CRAY-1 being a register-to-register machine (cf. section 2.2), whereas the CYBER 205 is a memory-to-memory computer (cf. section 2.3).
- Both machines are very popular and accessible for more and more users.

2.1 The Basic Computer Systems

The fundamental system suggested by John von Neumann in 1945 is a monoprocessor consisting mainly of five components:

Control Unit
Processor
Memory
Input Unit
Output Unit.

a)

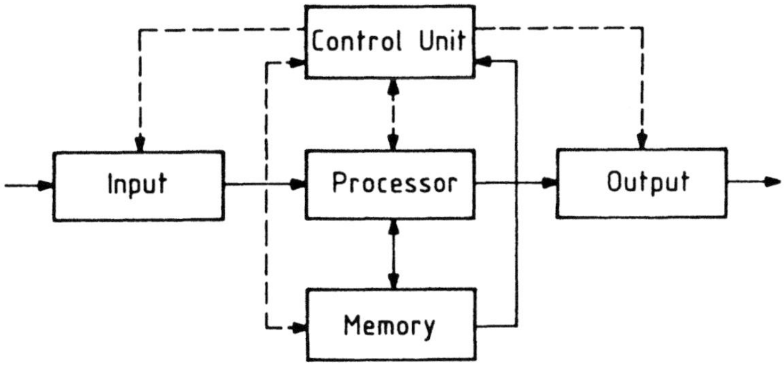

b)

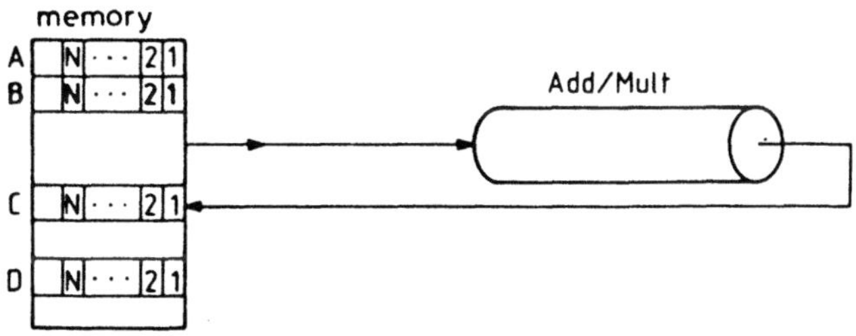

Fig. 2.1: a) Architectural block diagram of the principal units |90|, and b) operational scheme of a fundamental system |65|.

Fig. 2.1b shows the path of each pair of operands from the memory through the addition functional unit back to the memory. After the last result $a_{64}+b_{64}$ has reached the temporary array D, the multiplication of c_i with d_i starts again through the multiplication unit.

In our example D = (A + B) * C, we can give a rough estimate as to the number of cycles required for the computation of D. For simplicity, suppose that one arithmetic operation takes four clock cycles, and likewise for memory access. The cycle count for the fundamental system then includes the memory access of the first pair of operands (= 8 cycles), addition of 64 pairs of operands (= 256), storing the last result (= 4) - add and memory access resp. add and store may work in an over-lapped manner -, memory access to the intermediate results and the elements of C (= 8), multiplication of 64 pairs of operands (= 256) and finally storing the last result (= 4). The total cycle count for the example carried out on a von Neumann fundamental system is 536 clock cycles.

a)

b)

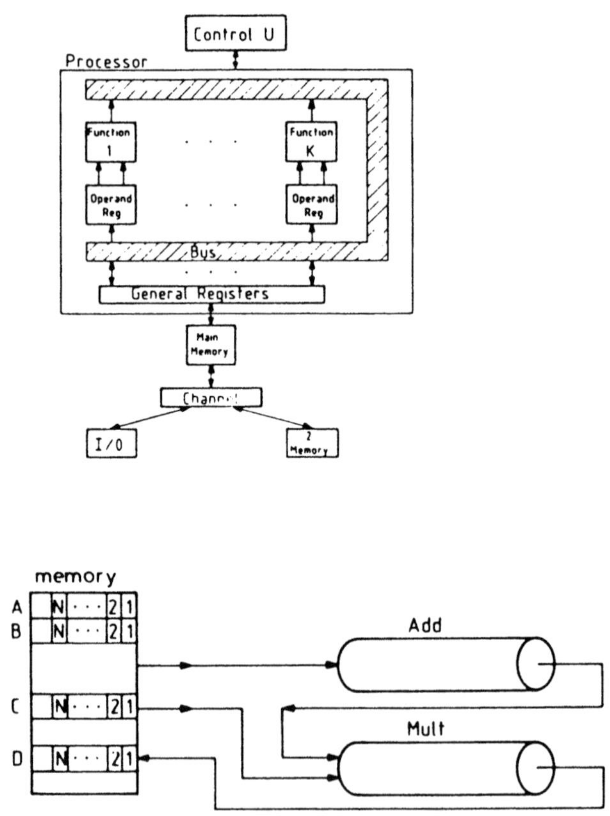

Fig. 2.2: a) Architectural block diagram of the principal units |90|, and b) operational scheme of a multi-function system |65|.

A bottle-neck in this system is the need to store temporary results back into the memory, although they still have to be processed again. The multi-function system overcomes this difficulty. It can work in parallel on different operations such as addition and multiplication, and is assisted by fast registers containing active data for the special purpose functional units. In our example, then, the results of the addition would not be stored back into the memory, but instead enter the multiplication functional unit, saving one store/fetch operation for each pair of operands, as illustrated in Fig. 2.2. The cycle count for the multi-function system then consists of the fetch operation of the first pair (= 8 cycles), addition of the first pair (= 4), multiplication of the results with the elements of C (= 256) and storing the last result. Over-lapped operational mode is assumed. The total number of clock cycles would then be 272.

a)

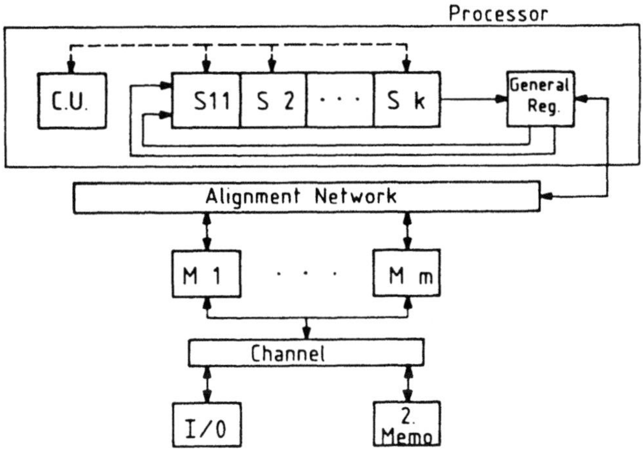

b)

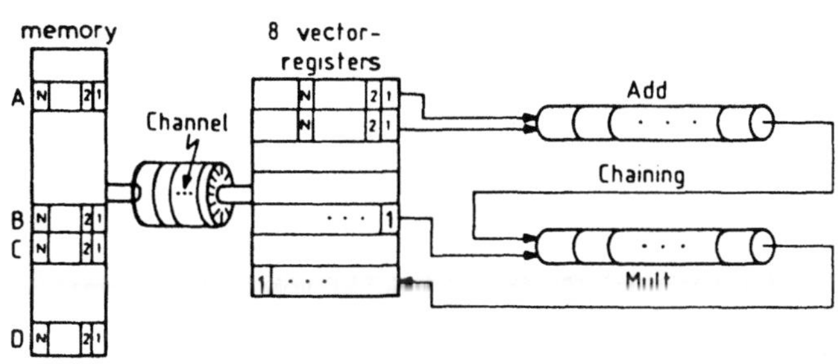

Fig. 2.3: a) Architectural block diagram of the principal units |90|, and b) operational scheme of a pipeline vector processor |65|.

The basic idea of pipelining on vector or pipeline processors is to segment the functional units into independent tasks S1,...,Sk (for simplicity, in our example here into four tasks). Then at each clock cycle a pair of operands enters the first segment of the pipe, while other pairs are in other stages, and at the other end one result is leaving the last segment simultaneously. Segmentation allows one result per clock cycle, which then enters the multiplication functional unit together with the corresponding element of the vector C. Fig. 2.3 shows the principle at work on a CRAY-1.

a)

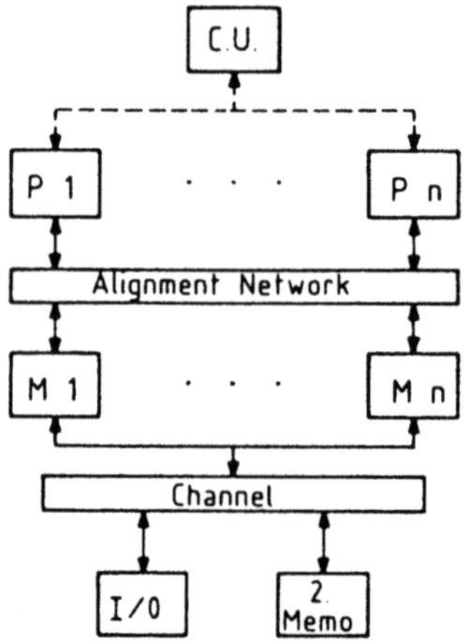

b)

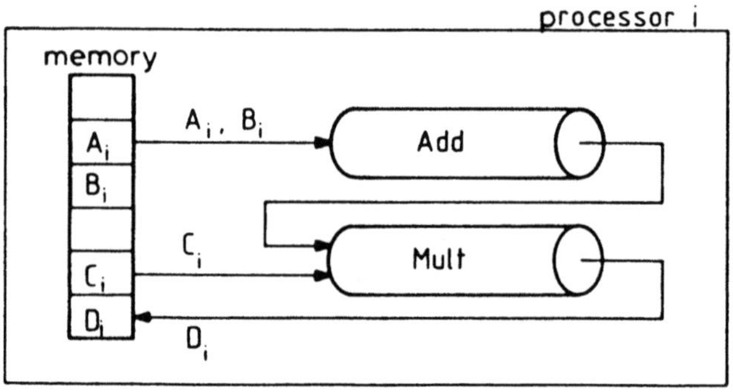

Fig. 2.4: a) Architectural block diagram of the principal units |90|, and b) operational scheme of a parallel processor |65|.

Again a cycle count for the example D = (A + B) * C results in 4 cycles for fetching the first element of A from the memory, 63 cycles for the remaining 63 elements of A to fill up the first vector register, 1 for the first element of B, 4 to fill the addition pipe, 4 to fill the multiplication pipe, 63 to multiply the remaining 63 pairs of operands and 4 cycles to store the last result back in the memory. A total of 143 clock cycles are all that would be required for the results.

The last system to be mentioned here is the parallel processor which consists of some or many single processors which in the case of a single instruction multiple data machine (SIMD, see |60|) are controlled by one master control unit (MCU, see |58|,|144|) and in the case of a multiple instruction multiple data machine (MIMD) are working independently on multiple data (see |99-101| and |189-191|). Fig.2.4 shows the working principle for one processor of a parallel SIMD machine.

To carry out a cycle count for a parallel processor we suppose that it consists of 64 single processing elements, each working on a single component of the vectors used in the calculation. It then needs 4 cycles to fetch the i'th element of A and 4 cycles to fetch the B_i, 4 to add and 8 to fetch and multiply corresponding elements. Finally 4 cycles are necessary to store back the i'th results. During execution all the processing elements are working in parallel. After 24 cycles the results are stored in the local memories of the processing elements.

For real computer configurations, the situation is of course much more complicated (see e.g. |14|, |53|, |72|, |77|, |78|, |85|, |88|, |90|, |103|, |104|, |118|, |122|, |123|, |144|, |153|); however, the principles indicated by the example are typical for the basic systems. The main lesson to be learnt here is that an improvement in efficiency can be due to the architecture of a computer and not necessarily only to the hard-ware.

2.2. The CRAY-1 Series

The first CRAY-1 was delivered to the Los Alamos Scientific Laboratory in 1976, only four years after its designer Seymour Cray left the Control Data

Corporation. It was the first commercially successful pipelined vector computer and to date more than 70 machines have been sold and installed.

The CRAY-1 provided 12 pipelined special-purpose functional units for the different arithmetic and logical operations, a fast clock period of 12.5 nanoseconds and a one-million word, bipolar memory with 50 nanoseconds cycle time, which is divided into 16 banks allowing fast access. Eight vector registers, each capable of holding 64 floating-point numbers, minimize main memory references and reduce bottle-necks due to insufficient main memory bandwidth.

The CRAY-1 is physically small with a diameter at floor level of about 3 metres (Fig. 2.5). The central cylindrical column is divided into three 90° segments. Each segment has four wedge-shaped columns holding up to 144 circuit modules, each of which comprises of a pair of circuit boards with the capacity for a maximum of two times 144 integrated circuit chips and approximately 300 resistor packages. The boards are based on a heavy copper heat transfer plate for the removal of the excess heat. The temperature of the copper plate is main-tained at 25° C by the circulation of Freon. The total power consumption of the machine is 128 KW.

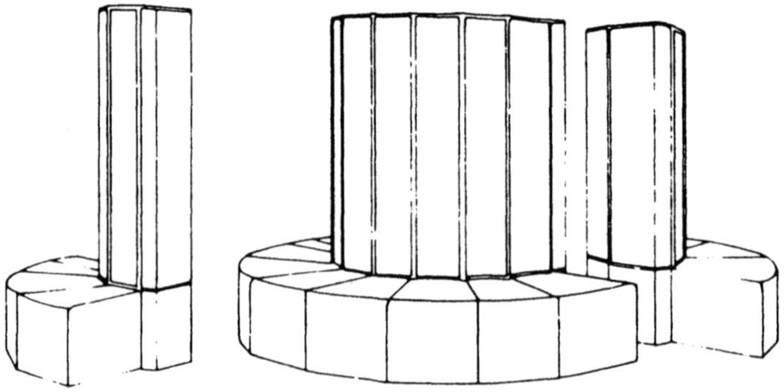

Fig. 2.5: A CRAY-1 or CRAY X-MP computer system with Input/Output Subsystem (IOS) and Solid-state Storage Device (SSD).

In April 1982 the CRAY X-MP system was announced which is an improved two-processor version of the CRAY-1. It offers a performance up to five times that of a CRAY-1 for certain vector routines. For production codes a factor of 1.5 per processor generally is realistic. The main differences between CRAY-1 and X-MP are the dual CPU's, four parallel memory access ports per processor and a 9.5 nanosecond clock cycle time. Data transfer rates to secondary storage can be improved by a factor of up to 100 over disks.

An Input/Output-Subsystem (IOS) completes both the CRAY-1S series and the CRAY X-MP. Two, three or four I/O-processors communicate with mass storage devices and with so-called front-end computers.

Finally, a Solid-state Storage Device (SSD) can be used as a fast-access disk. It consists of four columns arranged in a 90° arc (cf. Fig. 2.5), which matches the cabinet design of the CRAY-1S mainframe , and sets 64, 128 or 256 mega-bytes of storage at the programmer's disposal. With a CRAY-1S system with SSD, improvements in the transfer rate by a factor of up to 30 are possible.

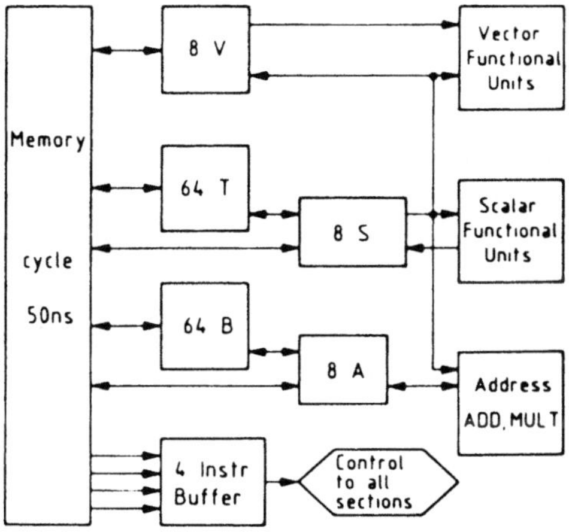

Fig. 2.6: Architectural block diagram of the principal units and data paths of the CRAY-1 |98|.

Knowledge of the mode of operation of such a machine is a prior condition for optimally vectorizing computer codes. We shall therefore describe the architecture of the CRAY-1S in more detail (see |49|, |97|, |98|).

The CRAY-1S computer consists of the central processing unit (CPU), the main memory with 0.5 or 1 million 64-bit words with single error correction, double error detection (SECDED) on each 64-bit word, expandable to 2 or 4 million words, and 12 input and 12 output channels. The memory contains bipolar 1024-bit or 4096-bit large-scale integrated (LSI) chips, arranged in 8 or 16 banks with a bank cycle time of four clock periods (50 nanoseconds).

Fig. 2.6 shows the basic functional areas of the CPU. Firstly, operands are fetched from main memory and stored in registers. The CRAY-1 has:

- 8 vector registers (V), each having 64 elements
- 8 scalar registers (S), each having 1 element
- 8 address registers (A), each containing 24 bits
- 64 temporary storage registers (T), each having 1 element
- 64 temporary storage registers (B), each containing 24 bits
- 4 instruction buffers with 64 16-bit registers each
- 12 (resp. 13 for the S-series) segmented functional units.

The primary operating registers are the scalar and vector registers. Scalar instructions perform functions such as addition or multiplication, by obtaining two operands from two scalar registers and entering one result into another scalar register. Analogous vector instructions repetitively perform the same function, obtaining new pairs of operands from elements of two vector registers during each clock period of 12.5 nanoseconds. Results are entered into elements of another vector register. The contents of the vector length register determine the number of operations performed by vector instructions.

The eight 24-bit A registers are primarily used as address registers for memory references, and as index registers, and also for shift count, loop control and I/O-operations. The address (resp. scalar) registers are supported by the 64 rapid-access temporary storage registers B (resp. T). The transfer rate between one A and one B register (resp. one S and one T register) is only one clock cycle.

All instructions are executed from four instruction buffers IP0, IP1, IP2 and IP3, each consisting of sixty-four, 16-bit registers in which substantial program segments can be stored. Forward and backward branching within the buffers is possible and program segments can be discontiguous.

In Table 2.1, the 12 (resp. 13 for the S series) fully segmented functional units handling the arithmetic, logical and shift operations are enumerated.

Each functional unit is independent of the others, and any number of functional units may be in operation simultaneously. A functional unit receives operands from registers and, when the function has been performed, delivers the result to a register.

Table 2.1: Functional units of the CRAY-1.

Functional Units	provide results to	clock periods for one operation
integer add	address registers	2
multiply		6
integer add	scalar registers	3
shift		2/3
logical		1
population count		3
integer add	vector registers	3
logical		2
shift		4
(population count)		2
floating point add	scalar and vector registers	6
floating point mult		7
reciprocal approx.		14

If the vector length is greater than 64 an internal vector loop is created similar to a DO-loop in FORTRAN having increment 64. For each vector segment of length 64, as well as for the remainder on division by 64, a vector instruction is generated which incurs a start-up time for reading the vector segment into the vector registers, (11 cycles, plus 63 cycles for the vector of length 64), and for filling the functional pipe. This start-up time is therefore dependent on the vector length, since a vector operation which, for example, may involve two vectors, can only be carried out when the first vector and the first element of the second vector have been read. Table 2.2 shows a rough cycle count for the addition of two vectors with vector length 100.

Table 2.2: Addition with vector length 100 on CRAY-1 (|64|, without cycles for vector length register, int.alia). Cycle time is 12.5 ns. Memory: V_1, V_2, V_3; Register: R_1, R_2, R_3.

Operation	Number of Cycles	Explanation
$V_1 \to R_1$	11	Access Time (AT) of the first element from V_1 to R_1
	35	reading of the next 35 elements from V_1
$V_2 \to R_2$	11	AT for first element from V_2 to R_2
	(35)	simultaneous reading and addition of the next 35 elements
$R_1 + R_2 \to R_3$	8	Functional Unit Time (FUT) for add-pipe
	35	addition of next 35 elements
$R_3 \to V_3$	10	AT of the first element to memory
	35	storage of the next 35 elements
$V_1 \to R_1$	11	AT of the 37th element from V_1 to R_1
	63	reading of the next 63 elements from V_1
$V_2 \to R_2$	11	AT of the 37th element from V_2
	(63)	simultaneous reading and addition
$R_1 + R_2 \to R_3$	8	FUT for the add pipe
	63	addition of the next 63 elements
$R_3 \to V_3$	10	AT of the first element from R_3 to the 37th element of V_3 into memory
	63	storage of the remaining 63 elements of R_3
Σ = 374 cycles = 4.675 seconds with 10^6 loops = 21 MFLOPS		

Because of the relatively short initiation time for vectors, vector operations with vectors having more than about five components are already faster in the vector mode than in the scalar mode. For long vectors, vector operations in the vector mode are approximately 5 to 15 times faster than in the scalar mode, in spite of the relatively large initiation and completion times: The registers must first be emptied before again a new 64 element segment can be fed into the active vector register. Under certain conditions, as for example in vector expressions of the form (A + B) * C, the result register of the first operation (here the addition) is used as the input register for the next operation (here the multiplication). The execution of the second operation begins as soon as the first element of the result vector arrives in the result register. This mode of operation is called "chaining". It saves the access time of the intermediate result A + B to the memory (see Fig. 2.3).

A programmer wanting optimal performance must try to re-use intermediate results as often as possible, filling and emptying vector registers while, if possible, useful work is performed in the functional units.

The rate of data transfer between the registers and main memory varies from one 64-bit word per clock cycle when the words are drawn from separate memory banks, to one word per four clock cycles when the words are drawn from the same memory bank. Successive memory addresses are allocated to successive memory banks, hence maintaining the maximum transfer rate, except in cases where the addresses are incremented by 8 or 16, where the rate is halved or quartered respectively (bank conflict). The data bus between the registers and main memory may be regarded as a pipe-line only 64 bits wide. The length of this pipeline is 11 clock cycles for scalars and 7 clock cycles for vectors.

The I/O-Subsystem of the CRAY-1S or CRAY X-MP consists of two to four I/O-processors and a buffer memory of up to eight million 64-bit words. The Master I/O-Processor coordinates the activities of the I/O-processors and controls the communication between the I/O Subsystem and the mainframe. It also manages the connections with the front-end computers.

The actual data transfer from main memory to I/O-Subsystem takes place on 1 or 2 high-speed channels each with a maximum transfer rate of 850 million bits per second. Transfer rate to a single magnetic disc is only 32 million bits per second, however a maximum of 10 disks (for 2 I/O-processors) or 18 disks (for 3 I/O-processors) can transfer simultaneously. Up to 46 disks can be configured on an I/O-Subsystem.

2.3 The CDC CYBER 205

The CYBER 205 manufactured by Control Data Corporation, is the result of a long program of research and developement, that began with the design of the CDC STAR-100 in 1965. The first STAR-100 was operational in 1973 and two machines were in service commercially in 1974/75.

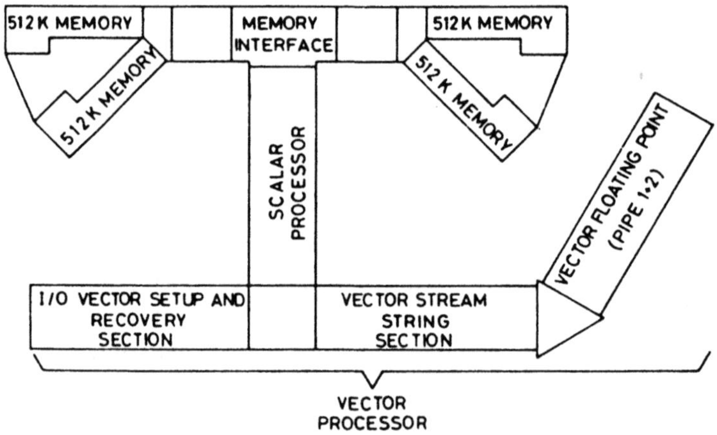

Fig. 2.7: Physical layout of a 2-pipe CYBER 205 with 2048K memory.

In 1973, Control Data began investigating large scale integration technology. Between 1974 and 1976 the design of a CDC 6400 class LSI test computer led to further study and development in packaging and cooling technology, to a corporate commitment in 1975 to redesign the STAR-100, and so to the CYBER 200 series of supercomputers.

The first member of this series was the CYBER 203 which was announced in

1979 and was tested in November 1979 (see |64|). It was made available to customers only as part of the CDC CYBER-NET services.

The second model, the CYBER 205, was announced in 1980, and the serial numbers 1 and 2 were tested as early as October and November 1980, respectively, |140|. Some technological highlights are its use of Large Scale Integration (LSI) circuitry throughout, high density chips with 168 gates each, improved packaging, subnanosecond switching time, dense integrated circuitry reducing interconnections, and a cooling system which maintains a low semiconductor junction temperature (For an explanation of these technical terms the reader is referred to the literature). The physical layout of the CPU and the central memory of the CYBER 205 is shown in Fig. 2.7 . The basic model of the CYBER 205 computer system consists of the central processing unit, one million 64-bit words of central memory with single-error correction and double-error detection on each 32-bit half word, 8 Input/Output ports, and a maintenance control unit. The CPU contains the scalar processor and a vector processor with one vector pipeline, both of which can operate in an overlapped manner.

The central memory capacity can be one, two, four or eight million 64-bit words. Moreover, virtual memory is supported by the operating system and hardware; the maximum size of the memory pages is 65K words corresponding to the 65K words maximum admissible vector length for vector operations.

The vector pipelines can be expanded to 2 or 4. Depending on the number of pipes 1, 2 or 4 results are generated per cycle, the cycle time being 20 nanoseconds. One type of operation can be executed at a time; the most important operations being distributed among the pipes in an even/odd fashion. Fig. 2.8 shows the basic functional areas of the CYBER 205.

The scalar arithmetic unit contains a sixty-four word discontiguous instruction stack segmented into eight super-words (SWORDs). The instruction stack is capable of holding up to 128 32-bit instructions, 64 64-bit instructions, or a combination of both, and provides a sixteen word instruction read ahead. The instruction issue pipe decodes all instructions, initiates scalar operations with the appropriate functional unit, and directs decoded vector/string instructions to the vector processor for execution.

The instruction issue pipe is capable of issuing instructions at the rate of one instruction every 20 nanoseconds. Thus, with independent vector and scalar instruction controls operating on a single instruction stream, the scalar processor can execute scalar instructions in parallel with most vector instructions provided there are no memory references generated by the scalar instructions.

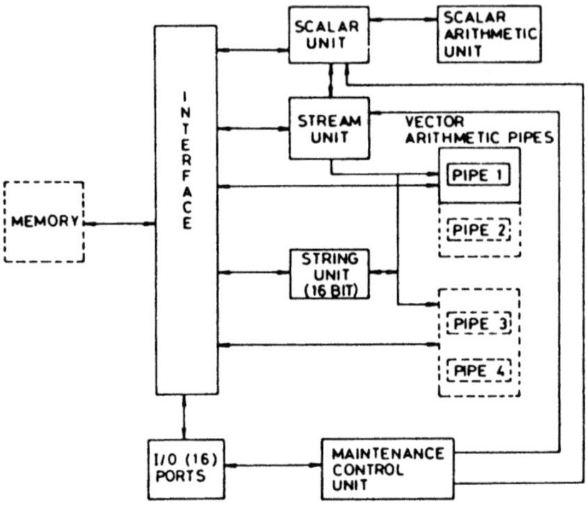

Fig. 2.8: Architectural block diagram of the principal units and data paths of the CYBER 205.

The scalar arithmetic unit contains completely independent functional units which are segmented and capable of accepting new operands every 20 nanoseconds. Except for the Divide/SQRT/Convert Unit, which must complete each operation before a new one can begin, all units are capable of being short-stopped. Short-stop is the process by which a result from any arithmetic unit may be returned directly to the input of any other arithmetic unit. This occurs in parallel with the storing of the result in the register file. Short-stop eliminates the time necessary to store the result in the register file and then retrieve it for use in the next arithmetic operation.

The scalar processor contains a semiconductor register file which provides 256 64-bit registers for use in instruction and operand addressing, indexing, field lengths, and as sources and destination registers for scalar instruction operands and result. The register file is capable of two reads and one write every 20 nanoseconds.

Table 2.3: Peak MFLOPS performance rates on the CYBER 205 with one, two and four pipes (one pipe for model 205-411 only).

NUMBER OF VECTOR PIPELINES		1	2	4
VECTOR ADD/SUBTRACT	32-BITS	100	200	400
	64-BITS	50	100	200
VECTOR MULTIPLY	32-BITS	100	200	400
	64-BITS	50	100	200
VECTOR LINKED MULTIPLY	32-BITS	200	400	800
AND ADD OR SUBTRACT	64-BITS	100	200	400
VECTOR DIVIDE/SQUARE	32-BITS	15.3	30.6	61.2
ROOT	64-BITS	8	16	32
VECTOR DIVIDE/SQUARE	32-BITS	-	61.2	122.4
ROOT (HIGH SPEED OPT.)	64-BITS	-	32	64

The vector processor consists of an operand streaming unit to control, buffer, and communicate data with the memory, a string unit for processing bit and byte operands, and general purpose segmented vector functional units, called vector pipelines, to process vector hardware instructions issued by the scalar unit. The vector processor is capable of manipulating vectors with 1-bit and 8-bit elements and 32-bit and 64-bit floating-point elements. Its high performance is achieved because of segmented operations synchronized with the basic 20 nanosecond clock period of the system and its method of streaming vector source and result operands directly from and to the main memory. In Table 2.3 some peak MFLOPS performance rates are given for the vector processor with one, two and four vector pipelines.

These asymptotic performance rates are, however, never attainable in reality. We therefore present the more realistic Table 2.4 which shows the MFLOPS rates of CYBER 205 vector add and multiply, depending on the vector length.

Table 2.4: Vector add and multiply MFLOPS rates on the CYBER 205.

VECTOR LENGTH	2 PIPE,64-BIT	4 PIPE,64-BIT 2 PIPE,32-BIT	4 PIPE,32-BIT
32	23.9	27.1	29.1
64	38.6	47.8	54.2
100	49.5	65.8	78.1
250	71.0	109.6	150.6
500	83.1	142.0	219.3
1000	90.7	166.1	284.1
10000	99.0	196.0	384.3
50000	99.8	199.2	396.8

A deeper insight into the operational scheme of the vector processor is necessary to vectorize computer programs successfully. In Fig. 2.9 it is shown how data are coming from the memory interface into the vector processor in three data input streams, two of which, A and B, are floating-point numbers and the third stream (M,X,Y) for control vectors and character strings. Two output streams C and R are for floating-point numbers and character strings respectively.

Each of the identical floating-point pipelines P1 to P4 comprises five separate segmented functional units for addition, multiplication, shifting operations and delay connected via a data interchange (see Fig. 2.10). Division and square root operations are performed in the multiplication unit. Each unit is attached to the data interchange by two input paths, A and B, and the output path C. If vector operations are of the special form

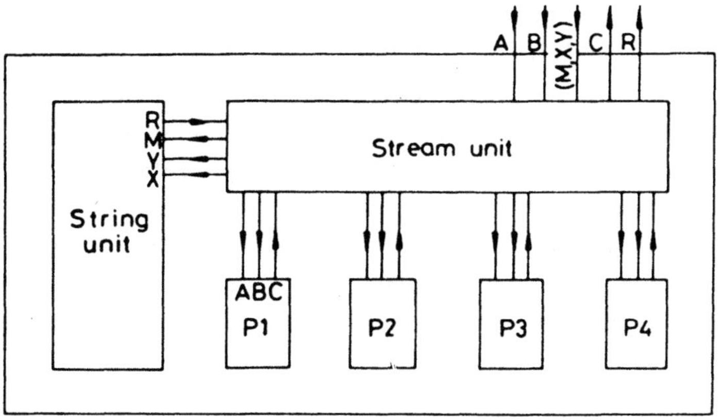

Fig.2.9: Architectural block diagram of the CYBER 205 vector unit |85|.

vector + scalar * vector
vector + scalar * scalar
scalar + vector * vector
scalar + scalar * vector ,

the two units "add" and "multiply" operate concurrently and the two instructions act as a single vector instruction with no intermediate reference to main memory. For such "linked triads" the performance of a floating-point pipeline is doubled.

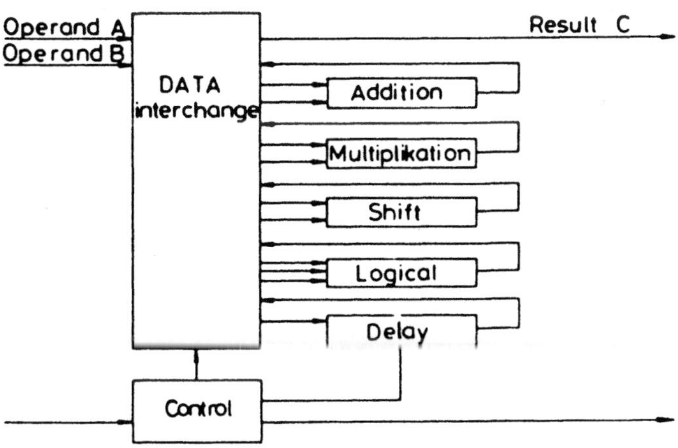

Fig.2.10: Structure of one CYBER 205 floating-point pipeline |85|.

As an example we discuss the organization of the addition unit in more detail (see Fig. 2.11). It is divided into seven segments. Around the add segment, a short-stop allows a non-normalized addition result of one element of a vector operation to be added to the next element. This facility is used in the interval instruction which forms the vector C(I+1) = C(I) + B, C(1) = A. Another short-stop returns the normalized result C to become the B operand input. The result arrives back at B eight clock periods after the operands involved entered the pipe, and hence one can accumulate C(I+8) = C(I) + A(I+8). This facility is used in summing all the elements of a vector, and in the dot or inner product instructions.

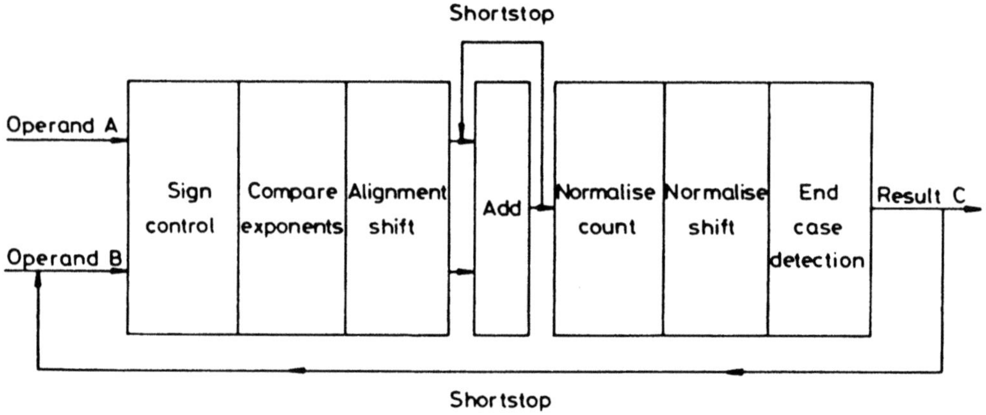

Fig. 2.11: Block diagram of the segmented floating-point addition unit of the CYBER 205 |85|.

3. Vectorization of FORTRAN Programs

In this chapter, we give an introductory survey of the vectorization of computer programs on the CRAY-1 and CYBER 205 vector computers. We place no great emphasis on good programming practices, nor do we present a complete survey of all possible ways to optimally vectorize production codes. Here we shall restrict ourselves to discussing some helpful coding techniques, which are easy to understand and to implement. However, it should be emphasized, that the selection of an appropriate algorithm is generally far more important than fancy programming techniques.

One point, though, is important: adapting FORTRAN codes to special architectures and to an optimizing compiler is not the invention of vector computer architects. It is well known that much faster execution of FORTRAN programs can be gained on serial computers if attention is paid to special programming practices (see e.g. |63|). A "checklist" for efficient programming especially with the IBM VS FORTRAN compiler is discussed by Scarborough |175|. It contains (among other items)

. Use unformatted in preference to formatted I/O
. Eliminate variably dimensioned arrays
. Pass subroutine arguments in common blocks
. Eliminate equivalence statements
. Insure that duplicate computations can be recognized
. Use arrays with identical dimension specifications
. Change division to multiplication where possible
. Avoid very short subroutines.

Some of them seem obvious, but some are more subtle. In |175| many examples underline the important points of this checklist.

3.1 Implementation of Serial Programs on Vector Computers

The production codes in any field of computational physics are usually highly complex. The vectorization of many thousands of statements will

Table 3.1: Section of a CYBER histogram.

address	frequency	relative time	
9480	5	.055	
9680	2	.055	
9880	2	.055	
9A80	4	.055	
9080	1	.055	
9E80	0	.055	
A080	1	.055	
A280	16	.056	
A480	6	.056	
A680	7	.056	
A880	10155	.269	vectorization
AA80	25349	.800	vectorization
AC80	5623	.918	vectorization
AE80	10	.918	
B080	4	.918	
B280	0	.918	
B480	12	.919	
B680	372	.926	
B880	995	.947	

Table 3.2: Flow Trace of the CRAY-1.

ROUTINE	TIME	%	CALLED	
1 FNH1R1	0.000395	0.00	1	
2 GRDFN	0.000061	0.00	21	
3 PUTZ	0.003958	0.04	61	
4 KEY	0.017824	0.17	1540	
5 PUTI	0.094767	0.90	8	
6 F	0.028835	0.27	16129	
7 PUTB	0.000375	0.00	1	
8 G	0.000915	0.01	512	
9 C	0.038148	0.36	21343	
10 CNFIX	o.058612	0.56	10	
11 RELAX	9.056721	86.16	370	← vectorization
12 RSCAL	0.323413	3.08	60	
13 INTAD1	0.888053	8.45	60	
*** TOTAL	10.512078			
*** OVERHEAD	1.121270			

naturally frighten every user; with the following strategy, however, an efficient vectorization of only parts of the program substantially improves efficiency, while not being particularly arduous:

STEP 1: Generation of a histogram showing the amount of CPU-time for different sections of the program. Table 3.1 shows a CYBER histogram for a multigrid solution of the Helmholtz equation. The Flow Trace of Table 3.2 demonstrates a similar situation on the CRAY-1. As is demonstrated, about 86 percent (related to 0.86 in the histogram) of the whole CPU time is spent in the subroutine RELAX solving the systems of linear equations with different mesh sizes. This is typical for most production codes arising in the numerical treatment of differential equations.

STEP 2: Hand-tailor the most time-consuming subroutines if auto-vectorization does not suffice. The CYBER and the CRAY compiler present, at the end of each subroutine, a complete list of the vectorized and non-vectorizable loops together with the reasons for non-vectorizability. The following example shows a non-vectorizable DO loop and the corresponding CYBER message (NN has not been specified) :

```
      .
      .
00171       DO  310    NL = 1, 100
00172       DO  210     N = 1, NN
00173  210  Z(N) = SCAL * X(N) + Z(N)
00174  310  CONTINUE
      .
      .
```

```
UNCOLLAPSABLE LOOPS              REASONS FOR NON-VECTORIZATION
     LINE 00171                  LINE 00172 LOOP WITH VARIABLE
                                 INITIAL/TERMINAL VALUE NESTED
                                 WITHIN LOOP
```

A nonlinear recursion is evaluated in DO loop 20 of the next example and the reason for non-vectorization is given for a CRAY:

```
10.          O(2)  = 4.
11.          DO 20   I = 3, M
12.          EL(I) = 1./O(I - 1)
13.          O(I)  = 4. - EL(I)
14.       20 CONTINUE
```

```
AT SEQUENCE NUMBER - 13.
PRNAME  LJAC        COMMENT- DEPENDENCY INVOLVING ARRAY "O"
                    IN SEQUENCE NUMBER 12
```

A number of microcoded hardware instructions and library routines is available for vector operations. The most important ones - for the CRAY and the CYBER - will be explained in the next two sections.

STEP 3: In the case of a highly serial algorithm involving, for example, linear and non-linear recurrences, step 2 is not successful and a complete restructuring of the algorithm is necessary. By way of illustration we consider in detail the influence of the remaining scalar parts of the program on the overall performance (see also |64|, |165|) :

Of a computer program with a total amount of z MFLOP (Million Floating Point Operations), y per cent of z are processed by the vector processor (VP) with a speed of V_V MFLOPS (MFLOP per Sec.), and the remaining x per cent with a speed of V_s MFLOPS on the scalar processor (SP). Let $V_V = a * V_s$, $a > 1$. Then one obtains the working time of the SP by

$$t_s = \frac{x}{100} \cdot \frac{z}{V_s} \text{ sec.}$$

The time which the VP uses for the y per cent is

$$t_V = \frac{y}{100} \cdot \frac{z}{V_V} \text{ sec}$$

If the SP and VP operate sequentially then the total CPU time of the job is

$$T_s = t_s + t_V.$$

On the other hand, if they operate strictly in parallel, there results

$$T_p = \max(t_s, t_v).$$

In practice neither a purely sequential nor a purely parallel operation of SP and VP occurs. One therefore would set more realistically:

$$T_j = (1-\alpha)T_p + \alpha T_s, \qquad 0 \leq \alpha \leq 1.$$

More interesting than the absolute CPU times are the average MFLOPS rates for the total program

$$V^s_{eff} = \frac{z}{T_s} = \frac{100\, a\, V_s}{ax + y}$$

for the purely serial SP and VP and

$$V^p_{eff} = \frac{z}{T_p} = \min\left\{\frac{V_s}{x}, \frac{V_v}{y}\right\} \cdot 100$$

for the parallel operation of SP and VP. For the general case of overlapped operations:

$$V_{eff} = \frac{100 \cdot V_s}{(1-\alpha) \cdot \max\left\{x, \frac{y}{a}\right\} + \alpha\left(x + \frac{y}{a}\right)}$$

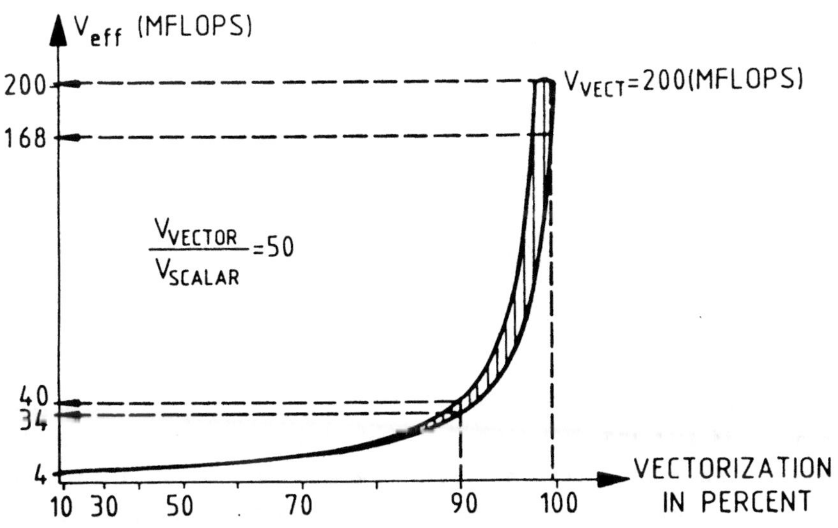

Fig. 3.1: MFLOPS rate diagram for example 1, upper curve $\alpha = 0$, lower curve $\alpha = 1$.

For a more detailed explanation of these considerations and the following diagrams we give some examples.

Example 1: The above mentioned multigrid solver is 90 per cent vectorized, i.e. y = 90, x = 10. For the following consideration we assume for simplicity that in the 90 per cent only linked triad instructions with large vector lengths are performed. Asymptotically this gives for 2-pipe CYBER 205 an MFLOPS-rate of 200 (see Table 2.3). The SP operates at 4 MFLOPS. From Figure 3.1 one sees that the average MFLOPS rate is only between V_{eff} = 34 and V_{eff} = 40 MFLOPS.

Example 2: For the CRAY-1 with approximately 40 MFLOPS for triads and the factor of 10 between SP and VP the average MFLOPS rate for example 1 lies between V_{eff} = 21 and V_{eff} = 40 MFLOPS, Fig. 3.2.

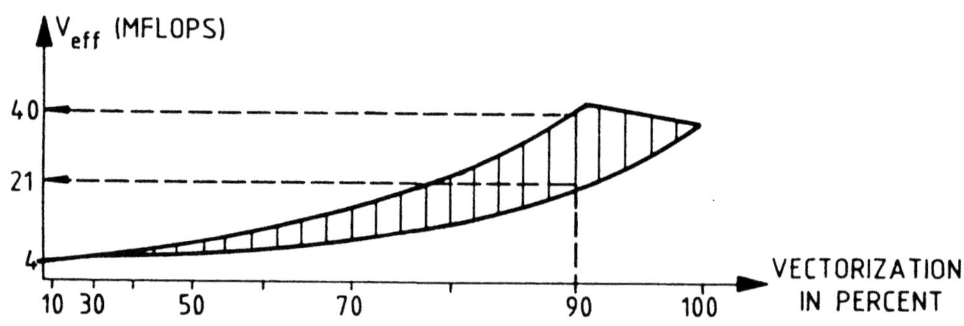

Fig. 3.2: MFLOPS rate diagram for example 2, upper curve $\alpha = 0$, lower curve $\alpha = 1$.

With full parallelism the CRAY operates, for z = 100 MFLOP, 2.25 sec. with a total of 44 MFLOPS and 0.25 sec. with 4 MFLOPS. The CYBER 205 on the other hand yields for 0.45 seconds a performance of 204 MFLOPS and for 2.05 seconds a performance of only 4 MFLOPS because of the influence of the non-vectorizable part of the program! Therefore on the CYBER 205 any further successful hand vectorization will lead to significant performance increases.

Example 3: Now let the portion of the VP in the example be y = 99 per cent. Then the average MFLOPS rate on the CYBER 205 according to Figure 3.3 amounts to between V_{eff} = 133 and V_{eff} = 202 MFLOPS. On the other hand if the program cannot be further improved, (y=90 per cent), then one must improve the performance of the SP five fold (V_s = 20 MFLOPS), in order to obtain MFLOPS rates of approximately 200.

These examples stress the importance of a good scalar processor performance of the vector machine just as much as that of a good autovectorizer.

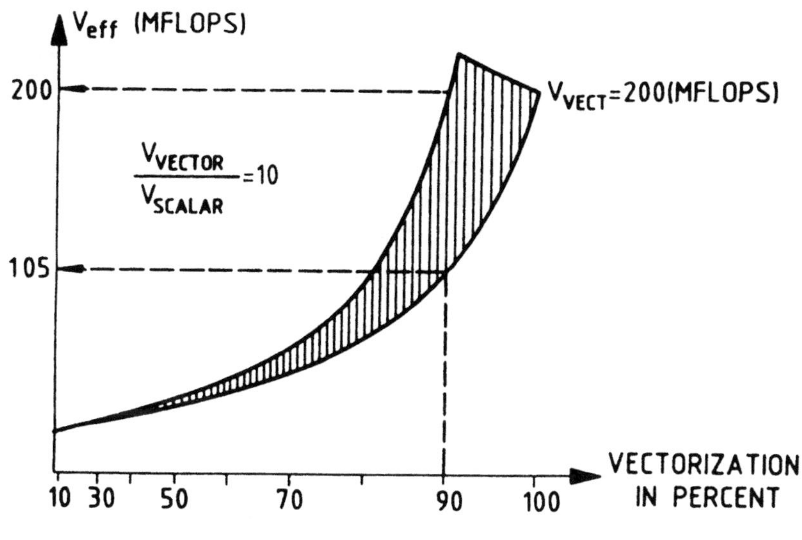

Fig. 3.3: MFLOPS rate diagram for example 3, upper curve $\alpha = 0$, lower curve $\alpha = 1$.

During the last decade, many production codes have been implemented on vector computers. In the literature of chapter 8.1, the interested reader may find some useful references dealing with the vectorization of various numerical algorithms.

3.2 Vectorization on the CRAY-1

The CRAY-1 is a register-to-register vector computer as described in section 2.2. As consequences concerning vectorization we discuss the following helpful techniques and their implementation in computer programs (see |84| for more details):

- Putting DO loops into subroutines or functions and vice versa
- Using few loops with long code blocks in preference to many short code loops
- Using long loops inside short loops rather than vice versa
- Special subroutines for linear recurrences
- Partial vectorization of irregular addressing
- Removing IF statements
- Manipulating operations so that they occur in an order that increases chaining.

The compiler only vectorizes innermost DO loops. Table 3.3 gives an impression of the vectorization possibilities (cf. |83|). For various syntactic constructions within innermost DO loops, Release 1.11 of the CRAY-1 FORTRAN Compiler (CFT) vectorizes all the "easy" and "straightforward" constructs.

Table 3.3: Classification of syntactic constructs with regard to vectorization on the CRAY-1 |83|.

Easy	- Long or complicated loops - Non unit incrementing of subscripts - Expressions in subscript - Intrinsic function references
Straightforward	- Scalar temporary variables - Function calls to programmer-supplied functions - Inner products - Logical IF statements - Reduction operations

Table 3.3 continued:

Difficult	- Linear recursions - Some IF statements - Complicated subscript expressions - Nonlinear indexing
"Impossible"	- Complicated branching within a loop - Ambiguous subscripting - Transfers into a loop - Subroutine calls - Nonlinear recursion - Some I/O

The FLOW TRACE option is used (see section 3.1) first to obtain a complete list of the subroutine calling tree and the time spent in each routine. One then normally starts vectorizing the most time consuming parts of the program.

As there are few FORTRAN extensions provided for the CRAY-FORTRAN compiler, most of the problems treated in this section deal with the restructuring of a sequence of standard FORTRAN statements. For example, subroutine and function calls within DO loops depending on the loop indices prevent the compiler vectorizing. The following sequence might have arisen in a program solving a finite difference equation (M1=M-1):

```
      DO    1   J = 2, M1
      DO    1   I = 2, M1
      PM = VELOC (I,J)
      PL = VELOC (I-1,J)
      PR = VELOC (I+1,J)
      CALL RELAX (PM, PL, PR)
      VELOC (I,J) = SQ2 (PM)
    1 CONTINUE
        .
        .
        .
```

```
      SUBROUTINE RELAX (PM, PL, PR)
      COMMON OM, HH
      PM = (1.-OM) * PM + 0.5 * OM * (HH + PL + PR)
      RETURN
      END

      FUNCTION SQ2 (P)
      DATA ALPHA  /.../
      SQ2 = ALPHA * SQRT (P)
      RETURN
      END
```

The innermost DO loop will not vectorize owing to the subroutine call and the call to a function not recognized by the compiler. Putting the loop inside the subroutine leads to:

```
        .
        .
        .
      CALL RELAXV (VELOC)
      CALL SQ2V (VELOC)
        .
        .
        .
      SUBROUTINE RELAXV (U)
      DIMENSION U(100, 100)
      COMMON OM, HH, M1
      OM1 = 1.-OM
      OM2 = 0.5 * OM
      DO   1   I = 2, M1
      DO   1   J = 2, M1
      U(I,J) = OM1 * U(I,J) + OM2 * (HH + U(I-1,J) + U(I+1,J))
    1 CONTINUE
      RETURN
      END

      SUBROUTINE SQ2V (V)
      DIMENSION V(100, 100)
      COMMON OM, HH, M1
      DATA ALPHA /.../
```

```
      DO   1   J = 2, M1
      DO   1   I = 2, M1
      V(I,J) = ALPHA * SQRT (V(I,J))
    1 CONTINUE
      RETURN
      END
```

Both subroutines will now vectorize. But in this example it would be better to put the subroutines inside the loop to increase the arithmetic in the inner loop:

```
       .
       .
       .
      DO   1    I = 2, M1
      DO   1    J = 2, M1
      VELOC (I,J) = OM1 * VELOC (I,J) + OM2 * (HH + VELOC (I-1,J) +
                    VELOC (I+1,J))
      VELOC (I,J) = ALPHA * SQRT (VELOC (I,J))
    1 CONTINUE
       .
       .
       .
```

In this example, moreover, with improved vectorizability, the program also achieves increased transparency.

The above instruction sequence is also an instructive example in the use of as few loops as possible, containing long code blocks, instead of many short-code vectorizable blocks. Consider the following sequence:

```
    CALL  VADD (A,B,C,N)
    CALL  VMULT(C,A,E,N)
    CALL  VADD (E,B,A,N)
```

Here one uses the vector subroutines VADD and VMULT. This version vectorizes, but the expanded combination

```
      DO  1   I = 1,N
      A(I) = (A(I) + B(I)) * A(I) + B(I)
    1 CONTINUE
```

is significantly faster than the series of calls. The sum A+B and the product (A+B)*A do not have to be stored, but can be kept in a register and A does not have to be fetched a second time. This is also an example of the manipulation of operations in order to increase chaining. Consider for example

```
      DO  1   I = 1,1000
      DO  1   J = 1,5
      A(I,J) = (A(I,J) + B(I,J)) * A(I,J) + B(I,J)
    1 CONTINUE
```

Since only the innermost DO loops are vectorized, the calculation with vectors of length 5 leads to a performance rate similar to that of scalar performance. Reversing the order of the I and J loops would cause an improvement factor of about 8 over the original code.

One of the most refractory problems on vector computers is the vectorisation of linear and non-linear recurrences. As this is more a question of algorithm, we shall return to it again later. For the moment however, we shall restrict ourselves to the implementation of single linear recurrences on the CRAY-1.

A linear recurrence uses the result of a previous pass through the loop as an operand for subsequent passes, and this prevents vectorization. An example of a first-order, linear recurrence is

```
      S(1) = A(1)
      DO  1   I = 1,N-1
      S(I+1) = -B(I) * S(I) + A(I+1)
    1 CONTINUE
```

A second-order, linear recurrence may be of the form

```
      S(1) = A(1)
      S(2) = A(2)
      DO    1  I = 1,N-2
      S(I+2) = B(I) * S(I+1) + A(I+2) * S(I)
    1 CONTINUE
```

In these cases straight forward vectorization is impossible. Therefore, CFT offers special subroutines which run with optimum efficiency on the CRAY-1, and which solve first-order, and some second-order, linear recurrences. The subroutine

```
    FOLR (N,A,INCA,B,INCB)
```

for example solves the above mentioned first-order linear recurrence. Here INCA and INCB are the skip distances between the elements of the vectors A and B, respectively. N is the length of the recurrence. The output overwrites the input vector B.

Even within more complex programs, vectorization of recurrences may still be straightforward (cf. section 5.4 for a more detailed discussion).

If the DO loop is not truly recursive, as for example in

```
      DO   1  I = 200, 300
      A(I) = A(I-L)
    1 CONTINUE
```

and L has some positive integer value between 101 and 200, the easiest approach is to try directing the compiler to vectorize the loop and see if the answers remain the same. The compiler directive

```
    CDIR$ IVDEP
```

placed immediately in front of the DO loop to be vectorized causes the computations to be performed in vector mode, provided that the loop contains no CALL or IF statements.

Another example of fictitious recursions often arises in problems with red-black and zebra-line structures (cf. sections 6.1 and 6.2) which can easily be vectorized by applying the same compiler directive:

```
          DO    40    J = 2,M1,2
    CDIR$ IVDEP
          DO    40    I = 2,M1,2
          U(I,J) = 0.25 * (U(I-1,J) + U(I+1,J) + U(I,J-1) + U(I,J+1))
       40 CONTINUE
```

For fixed J, the even subscripted values of U on the left hand side depend only on the odd subscripted ones on the right hand side, and the directive is appropriate.

In many applications no contiguous data structure is present. In the Monte-Carlo method (see section 9.5) we have to deal with randomly distributed data, while in three-dimensional problems it is necessary to gather and scatter two- and one-dimensional substructures. In FORTRAN this problem is expressed by subscripted subscripts as in the following example

```
      DO    1    I = 1, 100
      J = INDEX (I)
      A(I) = B(J) + ...
    1 CONTINUE
```

This loop can be partly vectorized by using a temporary array to first gather the irregularly distributed elements into a contiguous vector:

```
      DO    1    I = 1, 100
      J = INDEX (I)
    1 TEMP(I) = B(J)
      DO    2    I = 1, 100
    2 A(I) = TEMP (I) + ...
```

For problems with irregular addressing, gathering and scattering subroutines are available. The above example then simply reads as follows

```
      CALL   GATHER (100,TEMP,B,INDEX)
      DO     2   I = 1, 100
    2 A(I) = TEMP(I) + ...
```

The gather subroutine uses the integers pointed to by the array INDEX as indices to take possibly disparate elements from the vector B and make them contiguous in the vector TEMP.

As a last problem we deal with removing IF statements from innermost loops. Any loop containing an IF statement prevents the compiler from vectorizing. But, as is seen in Table 3.3, vectorization of some loops containing IF's may be straightforward while others are difficult but not impossible, depending on the structure of the code. Intrinsic functions described in the appendix of the CFT manual |36| may help to overcome some of these difficulties. The following example (see |83|)

```
      DO   1    I = 1, 100
      IF(A(I).LT.0.) A(I) = 0.
    1 B(I) = SQRT (A(I)) + ...
```

is already vectorized by the compiler (Release 1.11) and transformed into

```
      DO   1    I = 1, 100
      A(I) = AMAX1 (A(I), 0.)
    1 B(I) = SQRT (A(I)) + ...
```

which selects the maximum value of the two elements A(I) and 0.

In the next example, however, the user has to employ the vector merge operation CVMGT to merge the results of different vector computations (for example for inner and boundary points of a two-dimensional domain):

```
      DO   1   I = 1, 100
      IF(A(I).LT.0.)  GOTO 2
      B(I) = A(I) + C(I)
      GOTO 1
    2 B(I) = A(I) * C(I)
    1 CONTINUE
```

which can be converted to

```
      DO   1   I = 1, 100
      B(I) = CVMGT(A(I) * C(I),A(I) + C(I), A(I).LT.0.)
    1 CONTINUE
```

Some other vectorization aids are explained in more detail in |37|. A list of all scientific application subprograms available and a brief explanation may be found in the Library Reference Manual |36|. Subprograms are classified as follows:

- Basic linear algebra subprograms (BLAS)
- Other linear algebra subprograms
- Functions and linear recurrence routines
- Linpack
- Eispack
- Matrix inverse and multiplication routines
- Fast Fourier Transform routines
- Filter routines
- Gather and scatter routines
- Sort routines

BLAS is an important package of 22 single-loop level subprograms of hand-coded, assembly language routines that effectively use the pipeline/chaining properties of the CRAY-1 hardware wherever possible. A description of the functions along with a discussion of the algorithms used is given in |36|. Detailed discussions of the implementation of more complex problems and computer codes on the CRAY-1 may be found in |2|, |8|, |21|, |26|, |27|, |30|, |32|, |33|, |47|, |61|, |83|, |84|, |98|, |111|, |150|, |151|, |164|, |166|, |167|, |176|, |184|, |185|, |202| and |214|.

3.3 Vectorization on the CDC CYBER 205

The CYBER 205 is a memory-to-memory vector computer (cf. section 2.3). This architectural feature has a great influence upon vectorization.

The first point to note is that the CYBER 205 has a fixed start-up time that is required to set up the segmented general purpose pipelines for the vectors in question, i.e. to compute the first and last addresses for each vector, to transfer data between memory and arithmetic pipes, to initially fill the pipes, and various other overheads. This start-up time is 51 clock cycles (each 20 nanoseconds) for addition and multiplication. It does not depend on the vector length! Therefore performance increases as a function of vector length up to a length of $2^{16}-1$, the size of a large page in the memory. The longer the vector, the lower the portion of the start-up. Table 2.4 shows the performance rates sensitively depending on vector length.

Secondly, it is to be noted that the definition of a vector or array on a CYBER 205 is a contiguous set of memory locations. For real or integer vectors, the memory locations are words, for bit vectors they are bits. The need for long and contiguously stored vectors severely influences the process of vectorization. It is no longer sufficient to vectorize only innermost DO-loops which in practice often contain relatively short vectors of about 50 to 200 elements. In this case storing multi-dimensional arrays in one-dimensional, long and contiguous vectors (where possible) is more efficient.

For the CYBER 205, the range of problems from the easily vectorizable ones, to those for which vectorization is impossible is similar to that shown in Table 3.3. The main difference and difficulty for a programmer starting with vectorization on a CYBER 205 is the construction of loops containing long and contiguously stored vectors,using different vector instructions and vector functions (more than 200 special hardware instructions are available for the CYBER 205, but usually about 10 to 20 are important for the normal user).

In the remainder of this section, we shall not concern ourselves with the basic restrictions of the CYBER 205 such as

- vector length should be less than 2^{16} for a nest of loops,
- loops should not contain if, input, output or memory transfer statements or special functions and subroutines, and some others.

Instead, we concentrate on the construction of long and contiguously stored vectors (see |105 - 109|).

In the CDC vector language three syntax options are at the programmers disposal for treating vector operations on arrays, such as

```
      DO  1   I = 1,N
      C(I) = A(I) + B(I)
    1 CONTINUE
```

namely by

1. auto-vectorization
2. explicit descriptors: C(1;N) = A(1;N) + B(1;N)
3. implicit descriptors: CD = AD + BD.

Options 2 and 3 provide an opportunity to vector-process even if the auto-vectorizer does not recognize the vectors. A descriptor is a pointer to a vector. The internal format of a descriptor contains the starting address and the length of the vector pointed to. The explicit descriptor (see option 2) has the form

```
    A(L;N)
```

where L designates the vectors first element (in two dimensions for example 1,1) and N the length of the vector. The implicit descriptor consists of three parts. At first the specification statement

```
    DESCRIPTOR  AD
```

declares AD to be a descriptor. Before it is used, every descriptor is defined by a descriptor ASSIGN statement

```
    ASSIGN  AD, A(1;N)
```

The ASSIGN statement is an executable statement, so the vector pointed to by AD may be changed dynamically. Finally, the properly defined and initialized descriptors are used in arithmetic statements as in option 3 above. A more comprehensive example is the following outer product algorithm for matrix multiplication:

```
      DIMENSION  A(N,N), B(N,N), C(N,N)
      DESCRIPTOR AD, CD
      DO   1   J = 1,N
      ASSIGN  CD, C(1,J;N)
      CD = 0.0
      DO   1   K = 1,N
      ASSIGN AD, A(1,K;N)
      CD = CD + AD * B(K,J)
    1 CONTINUE
```

The algorithm will be explained in section 5.1 in more detail.

Let us now spend some time in reviving the popular example |107| of the addition of two-dimensional contiguous and non-contiguous arrays. Firstly, consider the following instruction sequence

```
      DIMENSION  A(N,N), B(N,N), C(N,N)
      . . .
      DO   1   J = 1,N
      DO   1   I = 1,N
      C(I,J) = A(I,J) + B(I,J)
    1 CONTINUE
```

The inner loop vectorizes, but only with vector length N. However the total data for each array is stored contiguously in one long vector in the memory:

A(1,1)	A(2,1)	...	A(N,1)	A(1,2)	...	A(N,2)	...	A(1,N)	...	A(N,N)

|←— I = 1 —→|←— I = 2 —→| |←— I = N —→|

which, in explicit descriptor notation, reduces to

```
    C(1,1;NN) = A(1,1;NN) + B(1,1;NN)
```

with NN = N * N. In implicit descriptor notation the example becomes

```
DIMENSION  A(N,N), B(N,N), C(N,N)
DESCRIPTOR AD, BD, CD
ASSIGN AD, A(1,1;NN)
ASSIGN BD, B(1,1;NN)
ASSIGN CD, C(1,1;NN)

CD = AD + BD
```

so that vector processing is performed with vector length NN = N * N. For N = 100, the resulting performance rates are (see Table 2.3)

Table 3.4: MFLOP-rates for two different vector lengths.

Vector length	MFLOPS 2-pipe	MFLOPS 4-pipe
N = 100	50	66.7
NN = 10 000	99	196.1

The long vector version executes 2 (resp. 3) times faster than the short vector version for a 2-pipe (resp. 4-pipe) CYBER 205.

In practice however, the active data are often non-contiguous in memory. Solving partial differential equations by finite difference or finite element methods results in procedures such as that exemplified by the following simplified instruction sequence

```
  DIMENSION  A(N,N), B(N,N), C(N,N)
  . . .
  DO   1   J = 2,N-1
  DO   1   I = 2,N-1
  C(I,J) = A(I,J) + B(I,J)
1 CONTINUE
```

The arithmetic operations take place only in inner grid points (I,J) of a two-dimensional domain. The boundary conditions are to be treated separately for I = 1, I = N and all J resp. for J = 1, J = N and all I.

Clearly the innermost DO loop vectorizes with vector length N-2. However, a further vectorization in this case is not straightforward. The total data elements are stored contiguously in the memory, but the active elements of each array are not:

A(2,2)	A(3,2)	...	A(N,2)	A(1,3)	A(2,3)	...	A(N,3)	A(1,4)	...	A(N-1,N-1)

|←—— J = 2 ——→|←—— J = 3 ——→| ...

Since each of the three arrays has the same active pattern (starting with the element (2,2))

ON	ON	...	ON	OFF	OFF	ON	...	ON	OFF	OFF	ON	...	ON

and most of data elements of the arrays are involved in the computation, the solution is to carry out the entire calculation and then store the results in the array C only for all I,J = 2,3,...,N-1. This is performed by a control bit vector having zeros as elements for the nonactive part of the arrays and ones otherwise. The corresponding in-line function

```
BITD = Q8VMKO (N-2,N; BITD)
```

produces the bit pattern

1,1,...,1,0,0,1,...,1,0,0,1,...,1

|←N-2→| |←N-2→| ...

Under the control of this "mask" the result of the addition is stored in the corresponding memory locations of the array C. The entire routine with this control storing technique reads as follows:

```
PARAMETER(N= ,N2=N-2,N12=N*N2)
DIMENSION  A(N,N), B(N,N), C(N,N)
DESCRIPTOR AD, BD, CD, BITD
BIT BITD, BIT (N12)
ASSIGN BITD, BIT(1;N12)
```

```
ASSIGN AD, A(2,2;N12)
ASSIGN BD, B(2,2;N12)
ASSIGN CD, C(2,2;N12)

BITD = Q8VMKO (N2,N;BITD)
WHERE (BITD)
CD = AD + BD
END WHERE
```

DIMENSION, DESCRIPTOR and BIT are declaration statements. The four assign statements associate each vector with a descriptor. The function Q8VMKO constructs the correct bit-mask for the arithmetic operation the result of which is stored in the array C only where BITD contains ones.

The WHERE block can also be replaced by the special call statement

```
CALL Q8ADDNV (,,AD,, BD, BITD, CD)
```

which directly generates a machine instruction. However the use of special calls (more than 240 different special calls exist in CYBER 205 FORTRAN) is not recommended for the average FORTRAN user. Special calls should only be used when absolutely necessary for specific programming tasks.

To avoid the descriptor declaration and assign statements one can use the explicit descriptor declaration which then reduces to

```
PARAMETER(N= ,N2=N-2,N12=N*N2)
DIMENSION  A(N,N), B(N,N), C(N,N)
BIT BIT(N12)
BIT(1;N12) = Q8VMKO(N2,N;BIT(1;N12))
WHERE (BIT(1;N12))
C(2,2;N12) = A(2,2;N12) + B(2,2;N12)
END WHERE
```

The bit vector controlled technique can also be applied to vectorize IF tests inside DO loops, where the arithmetic depends on the results of the IF test. Consider the following example

```
      DO   1   I = 1,N
      IF (B(I).LE.0.) THEN
         A(I) = B(I)
      ELSE
         A(I) = 1.-B(I)
      END IF
    1 CONTINUE
```

The calculation tree of this example is

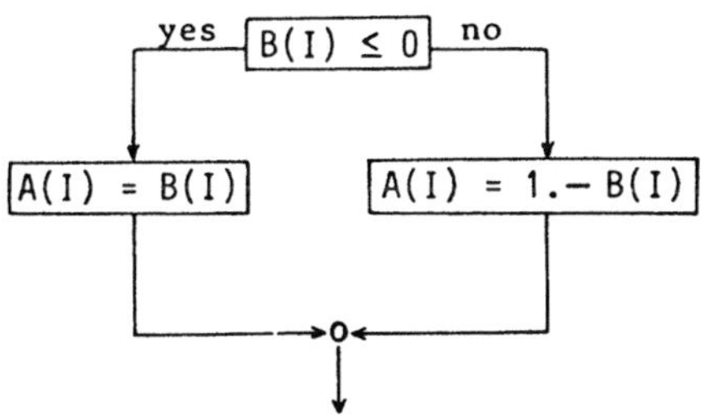

The first possible solution is to do all the arithmetic in both branches using the control bit vector to store the results:

```
BITD = BD.LE.0.
```

Using implicit descriptors, the vectorized form of the example then reads

```
DIMENSION  A(N), B(N)
DESCRIPTOR AD, BD, BITD
BIT BITD, BIT(N)
ASSIGN BITD, BIT (1;N)
ASSIGN AD, A(1;N)
ASSIGN BD, B(1;N)
BITD = BD.LE.0
AD = BD
WHERE (BITD)
AD = 1.-BD
END WHERE
```

If there are only a few 1's in the bit vector the vectorization is not very efficient since most of the arithmetic in the WHERE block is useless. In many problems it then would be better to compress the active elements into a contiguous vector. This is done by the COMPRESS statement:

```
CD = Q8VCMPRS (BD, BITD; CD)
```

which will cause those elements of the vector B that correspond to the 1's in the bit vector to be compressed into the vector C. The corresponding decompress operation which puts the answers back into the proper places of the array is

```
BD = Q8VXPND (CD, BITD; BD) .
```

The above example has the following form

```
BITD = BD.LE.0.
AD   = BD
CD   = Q8VCMPRS (BD, BITD; CD)
CD   = 1.-CD  ,
AD   = Q8VXPND (CD, BITD; AD)
```

Obviously if N is small, this is not very efficient because of the overhead caused by the compress and expand operations. If we want to eliminate several operands from a large number of N elements, this would take O(N) clock cycles to procure the corresponding operands, this being a high price to pay. For large N and many arithmetic operations on relatively long vectors however, it is clear that the COMPRESS will be more efficient. On the other hand, for large N but few active data elements, the SCATTER (resp. GATHER) statement overcomes such limitations and the price is only proportional to the number of those elements gathered. Suppose an integer data vector I(J), J = 1,2,..., M is given pointed to by ID which contains the M numbers of components involved in the calculation. Corresponding to the above example, a possible construction might be

```
J  = 0
DO   1    K = 1,N
IF (B(K).GT.0.)  GOTO 1
J  = J + 1
I(J) = K
```

```
1 CONTINUE
```

The statement

```
CD = Q8VGATHR (BD, ID; CD)
```

uses the numbers pointed to by ID as indices, to take different elements from B and make them contiguous in C. The corresponding inverse operation (scatter) puts the results of the calculation back into the proper places in the array.

The use of the scatter/gather approach is recommended if the output data elements are not in the same order as the input data (e.g. randomly distributed data) or if the active data is very sparse, e.g. in three-dimensional data structures, gathering and scattering two- or one-dimensional substructures. A further discussion of the use of control bit vectors, gather (scatter) and compress (merge) operations may be found in |10|.

In the meantime, many algorithms have been implemented on the STAR-100 and CYBER 205. The interested reader is referred to references |43|, |44|, |45|, |75|, |87|, |105|, |106|, |107|, |108|, |109|, |110|, |125|, |126|, |131|, |138|, |139|, |141|, |147|, |148|, |156|, |159|, |195| and |199|.

4. Vectorization of a Sample Program, on Different Vector and Parallel Computers

In this chapter, re-arranging and recoding of a sample program, (the numerical solution of the two-dimensional magnetohydrodynamic (MHD) differential equations), is discussed for the vector computers CRAY-1S, CDC-CYBER 205 and HITACHI's S9 with Integrated Array Processor (IAP). For comparison the same program is "parallelized" for the parallel computers ICL's Distributed Array Processor (DAP) and the Heterogeneous Element Processor (HEP) of DENELCOR. For a better understanding of the parallelized versions corresponding to the DAP and HEP, the interested reader requires a knowledge of the special FORTRAN-language versions (cf. |74|, |144| and |101|, |191|, respectively).

The MHD-program is a simple instructive example for an explicit finite difference code (see also |3|, |23|, |24|, |28|, |52|, |71|, |96|, |141|, |143|, |169|, |177|, |178| and |182| for the vectorization of finite difference algorithms), which is easy to vectorize and parallelize, a feature common to all explicit methods. Vectorization of the finite element method is explained in |3|, |19|, |40|, |50|, |95|, |127| and |155| in more detail. Consider a quadratic cross-section G with

$$G = \{(x,y) \mid 0 < x < 1,\ 0 < y < 1\}$$

and boundary $\dot{G}$. Under certain conditions (cf. |220|) the motion of the plasma through G may be described by the following model equations

$$\frac{\partial \underline{v}}{\partial t} + (\underline{v} \cdot \mathrm{grad})\underline{v} = -\mathrm{grad}\, p + \underline{j} \times \underline{B} \ ,$$

$$\frac{\partial \underline{B}}{\partial t} = \mathrm{rot}\,(\underline{v} \times \underline{B}) \ ,$$

$$\frac{\partial p}{\partial t} = -\underline{v} \cdot \mathrm{grad}\, p - p \cdot \mathrm{div}\, \underline{v} \ ,$$

$$\mathrm{div}\, \underline{B} = 0 .$$

where $\underline{j}$ = rot $\underline{B}$ and $\underline{B}$ means magnetic field, $\underline{v}$ velocity and p pressure of the plasma.

The steady-state solution of the above equations is of special interest. In this case, it has been shown in |220|, that the left hand side of the first equation can be replaced by $\underline{v}$ itself. Restriction to two dimensions results in

$$\underline{j} = \begin{pmatrix} 0 \\ 0 \\ JZ \end{pmatrix}, \quad \underline{B} = \begin{pmatrix} BX \\ BY \\ 0 \end{pmatrix}, \quad \underline{v} = \begin{pmatrix} VX \\ VY \\ 0 \end{pmatrix}.$$

The MHD-equations may now be rewritten as

$$JZ = \frac{\partial}{\partial x} BY - \frac{\partial}{\partial y} BX ,$$

$$VX = -\frac{\partial p}{\partial x} - BY \cdot JZ ,$$

$$VY = -\frac{\partial p}{\partial y} + BX \cdot JZ ,$$

$$\frac{\partial}{\partial t} BX = \frac{\partial}{\partial y} (VX \cdot BY - VY \cdot BX) ,$$

$$\frac{\partial}{\partial t} BY = -\frac{\partial}{\partial x} (VX \cdot BY - VY \cdot BX) ,$$

$$\frac{\partial}{\partial t} p = -(VX \frac{\partial p}{\partial x} + VY \cdot \frac{\partial p}{\partial y}) - p (\frac{\partial}{\partial x} VX + \frac{\partial}{\partial y} VY) .$$

The initial and boundary conditions are calculated using the stream function

$$\psi(x,y) = \frac{1}{4\pi^2} \sin \pi (x-1) \cdot \sin \pi (y-1) ,$$

which is an exact solution of the above system. More precisely, we get the initial conditions

$$p(x,y) = \pi^2 \cdot \psi^2 ,$$

$$BX(x,y) = -\frac{\partial \psi}{\partial y} ,$$

$$BY(x,y) = \frac{\partial \psi}{\partial x} ,$$

and for $t > 0$ the boundary conditions

$$\underline{B}_{(n)} = 0 \quad \text{and} \quad \underline{v}_{(n)} = 0$$

where (n) means normal component of the vectors $\underline{B}$ and $\underline{v}$, respectively. The velocities VX and VY in X- and Y-direction, the pressure P and the X- and Y-components BX and BY of the magnetic field are approximated in the (I,J)-th cell of a staggered grid as shown in Fig. 4.1.

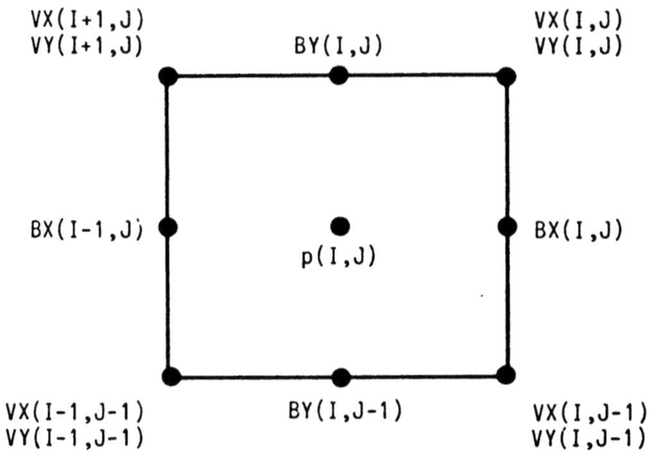

Fig. 4.1: Staggered grid for the discrete MHD-problem.

With that, one obtaines conservation of mass and momentum also for the discrete analogue.

For the remainder of this section, we discuss the implementation of the algorithm on a serial computer (e.g. IBM 3081). Then the main program written in FORTRAN IV is as follows (see |66|):

```
C       PROGRAM MAIN

        REAL BX(62,62),BY(62,62),VX(62,62),VY(62,62),
       *     P(62,62)
        COMMON /HAIN/ BX,BY,VX,VY,P
        COMMON /CONI/ N,N1,N2,L,N11,NN,LL
        COMMON /CONR/ SDT,DT,DX2,DX,EPS,SV2

        N=60
        LL=100
        EPS=0.0005
        DT=0.5
        N1=N+1
        N2=N+2
        NN=N*N
        N11=N1*N1
        DX=1./FLOAT(N)
        DX2=DX*DX

C       CALCULATION OF INITIAL VALUES

        PI=3.141592653590
        PS10=0.02533
        P1=PI*PS10
        P2=0.25*PS10
        D2=0.5*DX*PI
        DO      1       J=1,N2
        DO      1       I=1,N2
        X=(FLOAT(I-1)*DX-1.)*PI
        Y=(FLOAT(J-1)*DX-1.)*PI
        X2=X-D2
        Y2=Y-D2
        BX(I,J) =-P1*SIN(X)*COS(Y2)
        BY(I,J) = P1*COS(X2)*SIN(Y)
        P(I,J)=P2*(SIN(X2)*SIN(Y2))**2
      1 CONTINUE

        DO      2       J=1,N2
        DO      2       I=1,N2
        VX(I,J) = 0.0
        VY(I,J) = 0.0
      2 CONTINUE
        L=0
        SDT=0.
C
C       BEGIN OF THE ITERATION:
C       T=SECOND(T1)
        CALL TEXPL
C       T=SECOND(T1)-T
C
        WRITE
        STOP
        END
```

After storage has been reserved for the arrays BX, BY, VX, VY and P and calculation of parameters

NN	=	N * N = number of inner grid points
DX	=	Δx = grid spacing
DT	=	$\Delta t/\Delta x^2$ = Courant number, Δt = time step
EPS	=	required accuracy
LL	=	number of time steps

has been accomplished, initial conditions for BX, BY, P, VX and VY are evaluated. The main program finishes with a subroutine call to TEXPL, where all the arithmetic of the algorithm takes place:

```
      SUBROUTINE TEXPL
C
      COMMON /HAIN/ BX,BY,VX,VY,P
      COMMON /CONI/ N,N1,N2,L,N11,NN,LL
      COMMON /CONR/ SDT,DT,DX2,DX,EPS,SV2
      REAL BX(62,62),BY(62,62),VX(62,62),VY(62,62),
     *       P(62,62),DPX(62,62),DPY(62,62)
      REAL VB(62,62),ZJ(62,62),BXM(62,62),BYM(62,62),
     * VX1(62,62),VY1(62,62),DPX1(62,62),DPY1(62,62),
     * PX(62,62),PY(62,62)
C
      DT1=0.5*DT
      DT2=0.6258*DT1
   13 L=L+1
      SDT=SDT+DT*DX2
              DO        1         J=1,N1
              DO        1         I=1,N1
              PX(I,J)=P(I,J)+P(I,J+1)
              PY(I,J)=P(I,J)+P(I+1,J)
    1         CONTINUE

              DO        2         J=1,N1
              DO        2         I=1,N1
              DPX(I,J)=PX(I+1,J)-PX(I,J)
              DPY(I,J)=PY(I,J+1)-PY(I,J)
    2         CONTINUE
              DO        3         J=1,N1
              DO        3         I=1,N1
              BXM(I,J)=BX(I,J)+BX(I,J+1)
              BYM(I,J)=BY(I,J)+BY(I+1,J)
    3         CONTINUE
      DO        4         J=1,N1
      DO        4         I=1,N1
      ZJ(I,J)= BY(I+1,J)-BY(I,J)-BX(I,J+1)+BX(I,J)
    4 CONTINUE
      DO        5         J=1,N1
      DO        5         I=2,N
      VX(I,J)=(-DPX(I,J)-BYM(I,J)*ZJ(I,J))*0.5
    5 CONTINUE
```

```
      DO      6       J=2,N
      DO      6       I=1,N1
      VY(I,J)=(-DPY(I,J)+BXM(I,J)*ZJ(I,J))*0.5
    6 CONTINUE
      DO      7       J=1,N1
      DO      7       I=1,N1
      VB(I,J)=(VX(I,J)*BYM(I,J)-VY(I,J)*BXM(I,J))*DT1
    7 CONTINUE
            DO      8       J=2,N1
            DO      8       I=1,N1
            VX1(I,J)=VX(I,J)+VX(I,J-1)
            DPX1(I,J)=DPX(I,J)+DPX(I,J-1)
    8       CONTINUE
            DO      9       J=1,N1
            DO      9       I=2,N1
            VY1(I,J)= VY(I,J)+VY(I-1,J)
            DPY1(I,J)=DPY(I,J)+DPY(I-1,J)
    9       CONTINUE
      DO      10      J=2,N1
      DO      10      I=2,N1
      P(I,J)=P(I,J)*(1.-DT1*(VX1(I,J)-VX1(I-1,J)+
     *                         VY1(I,J)-VY1(I,J-1)))
     *                  -DT2*((VX1(I,J)+VX1(I-1,J))
     *                      *(DPX1(I,J)+DPX1(I-1,J))
     *                       +(VY1(I,J)+VY1(I,J-1))
     *                      *(DPY1(I,J)+DPY1(I,J-1)))
      BX(I,J)=BX(I,J)+(VB(I,J)-VB(I,J-1))
      BY(I,J)=BY(I,J)-(VB(I,J)-VB(I-1,J))
   10 CONTINUE

C     UPDATE OF BOUNDARY CONDITIONS

          DO      11      I=1,N1
          BX(I,1)  =BX(I,2)
          BX(I,N2)=BX(I,N1)
          BY(1,I)  =BY(2,I)
          BY(N2,I)=BY(N1,I)
          P(I,N2) =P(I,N1)
          P(N2,I) =P(N1,I)
          P(I,1)   =P(I,2)
          P(1,I)   =P(2,I)
   11     CONTINUE

C     ACCURACY TEST

      SV2=0.
      DO      12      J=2,N1
      DO      12      I=2,N1
      SV2=SV2+ABS(VX(I,J))
   12 CONTINUE
      SV2=SV2/NN
      IF(SV2.GE.EPS)  GOTO 13
C     IF(L.LT.LL)GOTO  13
      DT=2.*DT1
      RETURN
      END
```

In DO-loops 1 and 2, pressure derivatives are computed. The discrete analogue of the first differential equation is then evaluated by means of loop 4. In loops 5 and 6, evaluation of the X- and Y-component of the velocity takes place. In loops 7 to 10, the time derivatives of BX, BY and P are computed. The boundary conditions are re-evaluated in loop 11. Finally an accuracy test terminates the subroutine.

4.1 Vectorization on the CRAY-1S

CRAY FORTRAN is nothing more than standard FORTRAN, together with some special features (see section 3.2). We only have to remove the "C" in line 7 of the main program and the code runs successfully on a CRAY machine. In the subroutine TEXPL, a "possible vector dependency" arises in loop 11 where the arrays BX, BY and P appear on the left and on the right-hand side of the equation. However, they are not recursive, and by means of the compiler directive CDIR$ IVDEP, a vectorization of this loop will be forced. Writing the program in a more elegant form by breaking loop 10 into three or four smaller loops, is detrimental to the performance by a factor of up to 30 percent. The CRAY-compiler normally likes complicated innermost DO-loops, where "chaining" is possible (i.e. several functional units are working in parallel, see section 2.2). Finally we replace loop 12 by the CAL (CRAY Assembly Language) - routine SASUM. In addition, some minor changes such as loop unrolling might bring significant improvements (see section 3.2 and also CRAY's Optimization Guide |37|).

4.2 Vectorization on the CDC CYBER 205

As one might expect, vectorization of the sample program on the CYBER 205 is more complicated. The main reason for this is the need to convert two-dimensional arrays into one-dimensional arrays to get long and contiguously stored vectors. Therefore, a 62*62 array may be converted in a one-dimensional vector with length 3844, resulting in an improvement in performance by a factor of nearly 2.5 for the subroutine TEXPL. The main program is very similar to the original version. For programs with a large requirement in storage, it is sometimes useful to use the large page parameter RLP in the PROGRAM-statement:

PROGRAM MHD (OUTPUT, TAPE6=OUTPUT, RLP)

causing the dynamic stack to be put on a large page in memory. The dynamic stack is a part of the virtual memory in which temporary arrays are held during computation. In general this effects increased storage and shorter CPU-time.

Furthermore, minor changes concerning COMMON-blocks and DO-loops have been made. We now concentrate attention on the subroutine TEXPL:

```
      SUBROUTINE TEXPL
      PARAMETER (N2=62)
      COMMON /HAIN/ BX,BY,VX,VY,P,DPX,DPY,PX,PY
      COMMON /FLUR/ BXM,BYM,VB,ZJ,VX1,VY1,DPX1,DPY1
      COMMON /CONI/ N,N1,N11,N22,L
      COMMON /CONR/ SDT,SV2,DT,DX2,DX
      REAL BX(N2,N2),BY(N2,N2),VX(N2,N2),VY(N2,N2),P(N2,N2),
     *     DPX(N2,N2),DPY(N2,N2),PX(N2,N2),PY(N2,N2)
      DIMENSION W(N2),BXM(N2,N2),BYM(N2,N2),VB(N2,N2),
     *     ZJ(N2,N2),VX1(N2,N2),VY1(N2,N2),DPX1(N2,N2),DPY1(N2,N2)
      DESCRIPTOR BIT1D,BIT2D,BIT3D
      BIT BIT1D,BIT1(4000)
      BIT BIT2D,BIT2(4000)
      BIT BIT3D,BIT3(4000)

      N12=N1*N2
      L1 = N2 * N1 - 1
      L2 = N2 * N1 - 2
      L3 = N2 * N1 - 3
      L4 = N2 * N  - 1
      L5 = N2 * N  - 2
      L6 = N2 *(N-1) - 1
      DT1=0.5*DT
      DT2=0.0625*DT1
      ASSIGN BIT1D,BIT1(1;N12)
      BIT1D=Q8VMKO(N1,N2;BIT1D)
      ASSIGN BIT2D,BIT2(1;N12)
      BIT2D=Q8VMKO(N-1,N2;BIT2D)
      ASSIGN BIT3D,BIT3(1;N12)
      BIT3D=Q8VMKO(N,N2;BIT3D)
   13 L=L+1
      SDT=SDT+DT*DX2
```

```
WHERE (BIT1D)
  PX(1,1;L1) = P(1,1;L1) + P(1,2;L1)
  PY(1,1;L1) = P(1,1;L1) + P(2,1;L1)
  DPX(1,1;L1)= PX(2,1;L1) - PX(1,1;L1)
  DPY(1,1;L1)= PY(1,2;L1) - PY(1,1;L1)
  BXM(1,1;L1)= BX(1,1;L1) + BX(1,2;L1)
  BYM(1,1;L1)= BY(1,1;L1) + BY(2,1;L1)
  ZJ(1,1;L1) = BY(2,1;L1)-BY(1,1;L1)-B(1,2;L1)+BX(1,1;L1)
  VY(1,2;L6) = (-DPY(1,2;L6)+BXM(1,2;L6)*ZJ(1,2;L6))*0.5
END WHERE

WHERE (BIT2D)
  VX(2,1;L3)= (-DPX(2,1;L3)-BYM(2,1;L3)*ZJ(2,1;L3))*0.5
END WHERE

WHERE (BIT1D)
  VB(1,1;L1)=(VX(1,1;L1)*BYM(1,1;L1)-VY(1,1;L1)*BXM(1,1;L)
END WHERE

WHERE (BIT1D)
  VX1 (1,2;L4) = VX (1,2;L4) + VX (1,1;L4)
  DPX1(1,2;L4) = DPX(1,2;L4) + DPX(1,1;L4)
END WHERE

WHERE (BIT3D)
  VY1 (2,1;L2) = VY (2,1;L2) + VY (1,1;L2)
  DPY1(2,1;L2) = DPY(2,1;L2) + DPY(1,1;L2)
  P(2,2;L5) = P(2,2;L5)*(1.-DT1*(VX1(2,2;L5)-VX1(1,2;L5)
 1                             +VY1(2,2;L5)-VY1(2,1;L5)))
 2  - DT2*((VX1(2,2;L5)+VX1(1,2;L5))*(DPX1(2,2;L5)+DPX1(1,2;L5))
 3        +(VY1(2,2;L5)+VY1(2,1;L5))*(DPY1(2,2;L5)+DPY1(2,1;L5)))
  BX(2,2;L5) = BX(2,2;L5) + (VB(2,2;L5)-VB(2,1;L5))
  BY(2,2;L5)=BY(2,2;L5)-(VB(2,2;L5)-VB(1,2;L5))
END WHERE

  W(1;N1)=Q8VGATHP(BY(2,1;N12),N2,N1;W(1;N1))
  BY(1,1;N12)=Q8VSCATP(W(1:N1),N2,N1;BY(1,1;N12))
  W(1;N1)=Q8VGATHP( P(2,1;N12),N2,N1;W(1;N1))
  P(1,1;N12)=Q8VSCATP(W(1;N1),N2,N1; P(1,1;N12))
  W(1;N1)=Q8VGATHP(BY(N1,1;N12),N2,N1;W(1;N1))
  BY(N2,1;N12)=Q8VSCATP(W(1;N1),N2,N1;BY(N2,1;N12))
  W(1;N1)=Q8VGATHP( P(N1,1;N12),N2,N1;W(1;N1))
  P (N2,1;N12)=Q8VSCATP(W(1;N1),N2,N1; P(N2,1;N12))
  BX(1,1;N1)=BX(1,2;N1)
  BX(1,N2;N1)=BX(1,N1;N1)
  P(1,N2;N1)=P(1,N1;N1)
  P(1,1;N1)=P(1,2;N1)

  VX1(1,1;L5) = VABS(VX(2,2;L5);VX1(1,1;L5))
  SV2 = Q8SSUM(VX1(1,1;L5),BIT3D)
  SV2=SV2/N11
  IF(SV2.GE.EPS) GOTO13
  DT=2.*DT1
  RETURN
  END
```

Essential elements of the vectorized code are the BIT-vectors BIT1D, BIT2D and BIT3D (see also section 3.3) corresponding to selected points of the grid. The construction of the BIT-vectors is carried out by means of the hardware instruction Q8VMKO explained in section 3.3. In this program we use the explicit descriptor notation. Arithmetic takes place under the control of the BIT-vectors. Results are stored in memory only WHERE the corresponding element of the BIT-vector is a "1". In this example the use of WHERE-blocks instead of Q8VCTRL saves about 25 percent CPU-time.

For the calculation of the boundary conditions, since the distance between the elements is constant, we apply periodic GATHER/SCATTER, which has a shorter start-up time than the GATHER/SCATTER with the index-list Q8VINTL:

Q8VGATHP (A(1,1;L), N2,N1;X(1;N))

N2 : increment
N1 : number of gathered/scattered elements.

Finally, the absolute values of the elements of VX are stored in the vector VX1. The sum of all elements of VX1 is computed using Q8SSUM for the event of an accuracy test being necessary.

4.3 Vectorization on HITACHI's S9 with IAP

The HITACHI-S9 is a scalar machine with a standard and a high speed arithmetic unit (HSA). Its instruction set is compatible with the IBM-370. The corresponding machine for the Japanese domestic market is the M-280H with a different instruction set.

The Integrated Array Processor (IAP) is integrated in the sense that it shares memory and HSA with the S9. The IAP controls vector data and vector addresses and contains the micro-programs for the vector instructions, the HSA serving as a pipeline for the IAP. Standard arithmetic and HSA can work simultaneously. Similar to the CYBER 205, vector operands and results flow from memory to memory. One double

precision (64 bits) result is produced every two cycles with a start-up time of 45 cycles for add and multiply (cycle time: 30 ns). Long vectors are more important on the S9/IAP than they are on the CRAY, but less important than on CYBER, because the "effective cycle time" (time per result) is 60 ns on S9/IAP versus 10 ns on a 2-pipe CYBER 205, whereas start-up time is 23 "effective cycles" on S9/IAP and 104 "effictive cycles" on CYBER 205 for add or multiply. With this in mind programming as a whole is similar to programming on the CRAY; programming is done in "ordinary" FORTRAN (in fact the program was performed by using HITACHI's FORTRAN 77 - compiler), auto-vectorization, which is a compiler option, always refers to the innermost loop. The properties of the auto-vectorizers of S9/IAP and CRAY are not, however, the same in every respect; certain recursions, for instance, do vectorize on S9/IAP and do not on CRAY; on the other hand having a scalar variable on the left hand side of an indexed expression does not prevent vectorization on the CRAY but does on the S9/IAP.
The fact that only the innermost loop is a candidate for vectorization together with the necessity to work with long vectors forces the programmer to convert multi-dimensional arrays into one-dimensional arrays, and convert double DO-loops into single DO-loops, which in many cases is the same thing. As an example we include the recoding pattern and two examples of recoding of a simplified version of the MHD program which is described above.

The following changes in the organization of array operands in storage were made:

1. In the original source, the vector operands of the two-dimensional arrays (BX, BY, VX, VY, P, DPX, DPY, VB, ZJ, BXM, BYM, VXI, VYI, DPXI, DPYI, PX, PY) were organized in main storage as illustrated in Fig. 4.2 .
2. The following modifications have been made:
   In order to increase the vector length of each vector operation, each array was restructured into a one-dimensional array, and double DO-loops into single DO-loops. All the two-dimensional arrays (including BX, BY, P) were converted into one-dimensional arrays with a length of (N+1)*(N+2) as shown below.

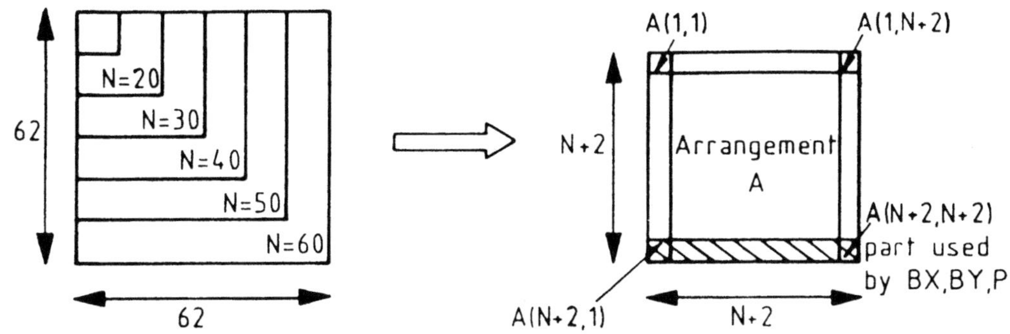

Fig. 4.2 : Storage mode of array A.

The elements "N+2,J"(J=1 to N+2) of the array are not used except by BX, BY, P. Vectors BX(N+2,J), BY(N+2,J) and P(N+2,J) (J=1 to N+2) are redefined as SBY, SBX and SP.

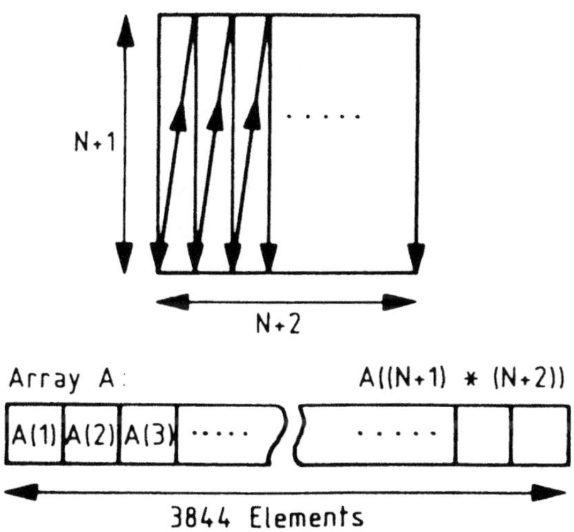

Fig. 4.3 : One-dimensional storage mode of a two-dimensional array.

An area of 62x62(=3844) has been allocated, though 62x61(=3782) would have been enough. We now illustrate the conversion of two-dimensional arrays into one-dimensional arrays by means of an example:

Original code:

```
      N1 = N + 1
      DO  6 J = 2,N
      DO  6 I = 1,N
      VY(I,J) = (BXM(I,J)*ZJ(I,J)-DPY(I,J)) * 0.5
    6 CONTINUE
```

Modified code:

```
      N1  = N  +  1
      N11 = N1 * N1
      DO  6  I = N1 + 1,N11 - N1
      VY(I) = (BXM(I)*ZJ(I) - DPY(I)) * 0.5
    6 CONTINUE
```

4.4 Vectorization on the ICL DAP

The DAP (Distributed Array Processor) is a "processor-matrix" of 64x64 1-bit-processing elements (PE) that works in a lock-step fashion as an SIMD-machine (Single Instruction Multiple Data stream), the instructions being controlled and issued by a master control unit. Each PE has direct communication to its four neighbours. More general communication can be realized by transmitting data over row or collumn highways. Sequential parts of the program run on the host (ICL-2900 series). The programmer explicitly has to devide his program into a host part and a set of subroutines that have to run on the DAP. Programming the DAP is done with an extended FORTRAN language, which contains constructs for describing vectors and matrices similar in philosophy to that in the programming language APL.

From the host's point of view the DAP can be thought of as being a 2 MByte storage module (4 KBit storage for each PE). This is in fact the way data are loaded into the DAP. There are two formats in which data is held in DAP storage. In the "vertical" or "matrix" format each number is held entirely in each PE. In the "horizontal" or "vector" format each operand is spread along a row of PEs. Obviously a considerable performance of the DAP can only be achieved if the programmer is successful in mapping his application onto the 64x64-world, keeping communication among PEs as local as possible.

To use the DAP for the MHD-program, suitable parts of the program need to be written in DAP-FORTRAN. The main program of MHD remains unchanged, though one must remove the "C" in line 7. Because of the relatively slow host computer, an ICL-2980 of the Queen Mary College in London, a DAP-module INIT for the calculation of the initial conditions has been built. The two DAP-modules then become:

```
      ENTRY SUBROUTINE TEXPL 3
      COMMON /HAIN/ VX,VY,P
      COMMON /CONI/ NX,NX1,NX2,L,NX11
      COMMON /CONR/ SDT,DT,DX2,DX,EPS,SV2,DT1,DT2
      COMMON /ANFA/ BX(,),BY(,),DPX(,),DPY(,),DPX1(,),DPY1(,),
     *              VX1(,),VY1(,),BXM(,),BYM(,),VB(,),ZJ(,),
     *              PX(,),PY(,)
      REAL VX(,),VY(,),P(,)
      REAL SDT,SV2,DT,DX2,DX,DT1,DT2
      LOGICAL LMAT1(,),LMAT2(,),LMAT3(,),LMAT4(,),LMAT5(,),LMAT6(,)
C
C     CONVERSION OF 2900 TO DAP
C
      CALL CONVFSI(NX,5)
      CALL CONVFSE(SDT,7)
      CALL CONVFME(VX)
      CALL CONVFME(VY)
      CALL CONVFME(P)
C
C     CONSTRUCTION OF LOGICAL MATRICES
C
C
C
      LMAT1=ROWS(1,NX1).AND.COLS(1,NX1)
      LMAT2=ROWS(2,NX).AND.COLS(1,NX1)
      LMAT3=ROWS(1,NX1).AND.COLS(2,NX)
      LMAT4=ROWS(2,NX1).AND.COLS(2,NX1)
      LMAT5=ROWS(1,NX1).AND.COLS(2,NX1)
      LMAT6=ROWS(2,NX1).AND.COLS(1,NX1)
C
C
      CALL INIT
C
   13 L=L+1
      SDT=SDT+DT*DX2
C
      PX(LMAT1) = P+P(,+)
      PY(LMAT1) = P+P(+,)
      DPX(LMAT1)= PX(+,)-PX
      DPY(LMAT1)= PY(,+)-PY
C
```

```
      BXM(LMAT1)= BX+BX(,+)
      BYM(LMAT1)= BY+BY(+,)
      ZJ(LMAT1) = BY(+,)-BY+BX-BX(,+)
C
      VX(LMAT2) = (DPX+BYM*ZJ)*(-0.5)
      VY(LMAT3) = (DPY-BXM*ZJ)*(-0.5)
      VB(LMAT1) = (VX*BYM-VY*BXM)*DT1
C
      VX1(LMAT5)= VX+VX(,-)
      VY1(LMAT6)= VY+VY(-,)
      DPX1(LMAT5) = DPX+DPX(,-)
      DPY1(LMAT6) = DPY+DPY(-,)
C
      P(LMAT4) = P*(1.-DT1*(VX1-VX1(-,)+VY1-VY1(,-)))
     *           -DT2*((VX1+VX1(-,))*(DPX1+DPX1(-,))
     *                +(VY1+VY1(,-))*(DPY1+DPY1(,-)))
      BX(LMAT4)= BX+(VB-VB(,-))
      BY(LMAT4)= BY-(VB-VB(-,))
C
      BX(,1)   = BX(,2)
      BX(,NX2) = BX(,NX1)
      BY(1,)   = BY(2,)
      BY(NX2,) = BY(NX1,)
      P(,NX2)  = P(,NX1)
      P(NX2,)  = P(NX1,)
      P(,1)    = P(,2)
      P(1,)    = P(2,)
C
      ZJ   = ABS(VX)
      SV2  = SUM(ZJ)
      SV2  = SV2/NX11
      IF(SV2.GE.EPS)  GOTO 13
C
      CALL CONVSFI(NX,5)
      CALL CONVSFE(SDT,5)
      CALL CONVMFE(VX)
      CALL CONVMFE(VY)
      CALL CONVMFE(P)
C
      RETURN
      END
C
      SUBROUTINE INIT
C
      COMMON /HAIN/ VX,VY,P
      COMMON /ANFA/ BX(,),BY(,),DPX(,),DPY(,),DPX1(,),DPY1(,),
     *              VX1(,),VY1(,),BXM(,),BYM(,),VB(,),ZJ(,),
     *              PX(,),PY(,)
      COMMON /CONR/ SDT,DT,DX2,DX
      INTEGER V(),PLACE
      REAL V1(),V2(),VX(,),P(,),VY(,)
C
      V=0
      PI=3.1415927
      PSI0=0.02533
      PSI1=PSI0*PI
```

```
C
C     MATRIX V WITH ELEMENTS FROM 0 TO 63
C
      PLACE=1
      DO      1       K=1,6
      V(ALT(PLACE))=V+PLACE
    1 PLACE=PLACE*2
C
      V1=PI*(EFLOAT(V)*DX-1.)
      V2=V1-0.5*DX*PI
      BX=-PSI1*MATC(SIN(V1))*MATR(COS(V2))
      BY= PSI1*MATC(COS(V2))*MATR(SIN(V1))
      P  =0.25*PSI0*MATC(SIN(V2)**2)*MATR(SIN(V2)**2)
C
      RETURN
      END
```

Here we give only a cursory description of these modules. A fuller discussion of the DAP FORTRAN language may be found in Parkinson |144|, |148|, Gostick |74|, the DAP Technical Reports or the DAP FORTRAN Manual.

A DAP FORTRAN-program cannot obtain direct access to data in FORTRAN storage mode. Consequently data must first be converted, within the host section, to either scalar, vector, or matrix storage mode. Similarly, when control is returned from the DAP section to the host section, it must first be converted to FORTRAN storage mode as done at the beginning and at the end of the subroutine TEXPL using the conversion subroutines CONVXXX. For example, the subroutine CONVFME converts data in FORTRAN storage mode (F) to matrix storage mode (M), and the data being converted being of type REAL (E).

The next step is the construction of logical arrays under the control of which the calculated data must be stored. This is similar to the BIT-vector controled store on the CYBER 205. As can be seen in the above example, the clear DAP FORTRAN remarkably simplifies programming.

The implementation of various algorithms on the DAP is discussed in detail e.g. in references |4|, |8|, |58|, |74|, |93|, |94|, |95|, |135|, |137|, |144|, |145|, |146|, |154| and |155|.

4.5 Parallelization on DENELCOR's HEP

The HEP is a machine with a shared resource MIMD architecture (Multiple Instruction Multiple Data stream,see section 2.1). Up to 16 Process Execution Modules (PEM) can be connected to up to 128 data memory moduls via a pipelined switch with a maximum capacity of 10 Mwords/sec per PEM.

Each data memory modul can contain up to one million words with a word length of 64 bits. A maximum number of 50 processes determined by the programmer can be alive in each PEM. Eight totally independent processes can be executing simultaneously, where execution of instructions can be visualized as being performed by an 8-stage pipeline. A new instruction can enter the pipeline every 100 nanoseconds. If access to the memory is made, the corresponding process is assigned to a queue that provides input to the pipeline switch. Only after the access to the memory has been completed, the process is re-assigned to the execution queue.

Processes within a common protection domain (task) can communicate with each other. Synchronization between processes is performed in a producer/consumer fashion. Typically, an instruction tests its sources for the state 'full' and its destination for the state 'empty'. If any of these tests fail, the corresponding process must wait until its requirements are met.

If all tests succeed, the instruction sets its sources to 'empty' and its destination to 'reserved'. When the function unit eventually writes a result into the destination location of the instruction, it sets the state of the destination to 'full'.

Even though each PEM is capable of performing 10 MIPS, one has to realize that in order to perform arithmetic operations on arrays, at least two additional instructions (branch and index increment) are necessary for a floating-point operation cycle to be completed. As a rule of thumb for simple arithmetic operations on indexed operands in a loop, one should not expect much more than 3 MFLOPS per PEM.

Despite of the rather extraordinary architecture, the FORTRAN programmer only must cope with a surprisingly small number of FORTRAN extensions (of course, as examples show, the complexity of a game is not proportional to the number of its rules). A process is created by the CREATE command that works syntactically entirely analogous to the CALL of a subroutine. Processes, in fact, are always performed in subroutines, except for the process which constitutes the main program. Especially marked variables, the so called asynchronous variables that have a state in addition to their value, and a few intrinsic functions that can test or modify the state of an asynchronous variable, complete the set of syntactical novelties with which the FORTRAN programmer must aquaint himself. The handling of the more strategic aspects of programming on the HEP is by far less straight forward: the created overhead must be kept low compared to the work done by the created process ("parallelizing the outer loop") and closely connected with that, the synchronization must be performed efficiently (and correctly). A considerable number of synchronization techniques have already been published in the user manual and the literature (cf. |99|, |100|, |101|, |189|, |190| and |191|).

The main tool for a programmer to enforce parallel execution and interaction of processes, is the class of asynchronous variables, each of which is associated with a full/empty state. HEP-hardware support allows the reading of such a variable only when full, after which it is set to "empty", and prevents writing until it is empty, with associated change of state to "full". These variables are distinguished from ordinary variables by names beginning with a $ symbol . If such a variable appears on the right hand side of an assignment statement, the meaning is "wait for full, read and set empty"; if it appears on the left side of the assignment, the meaning is "wait for empty, write and set full". This technique will now be illustrated using the program MHD.

In the subroutine TEXPL we distinguish between three main parts:

1. DO-loops 1 to 9
2. DO-loop 10
3. DO-loops 11 and 12 ,

where in each of these parts NPROC processes can work in parallel. If one process has finished its work for a certain index J and there is more work to be done, it automatically choses the next index J which has not yet been

processed. To demonstrate this, we present the re-coded version of TEXPL:

```
      SUBROUTINE TEXPL
      PARAMETER (IZ=62)

      REAL BX(IZ,IZ),BY(IZ,IZ),VX(IZ,IZ),VY(IZ,IZ),
     *     P(IZ,IZ),DPX(IZ,IZ),DPY(IZ,IZ)
      REAL VB(IZ,IZ),ZJ(IZ,IZ),BXM(IZ,IZ),BYM(IZ,IZ)
      REGISTER REAL DT3
      COMMON /HAIN/ BX,BY,VX,VY,P
      COMMON /CONI/ N,N1,N2,L,N11,NN,LL
      COMMON /CONR/ SDT,DT,DX2,DX,EPS,SV2
      COMMON /MHDP/ DPX,DPY,VB
      COMMON /MHD1/ DT1,DT2
      COMMON /MHD2/ BXM,BYM,ZJ
      LOGICAL $FIN,F
      REGISTER INTEGER I,J
      NPROC=N1
      IF(N1.GT.25) NPROC=25
      NPROC1=NPROC-1
      DT1=0.5*DT
      DT2=0.06258*DT1
   13 L=L+1
      SDT=SDT+DT*DX2

      PURGE $J,$JW,$FIN
      $J=1
      $JW=0
      DO      2       J=1,NPROC1
    2 CREATE OPB($J,$JW,$FIN,NPROC)
      CALL OPB($J,$JW,$FIN,NPROC)
      F=$FIN

      PURGE $J,$JW,$FIN
      §J=2
      $JW=0
      DO      15      J=2,NPROC1
   15 CREATE OPA($J,$JW,$FIN,NPROC1)
      CALL OPA($J,$JW,$FIN,NPROC1)
      F=$FIN

      PURGE $J,$JW,$FIN,$SV1
      $J=1
      $JW=0
      $SV1=0.0
      DO      16      I=1,NPROC1
      CREATE OPC($J,$JW,$FIN,NPROC,$SV1)
   16 CONTINUE
      CALL OPC($J,$JW,$FIN,NPROC,$SV1)
      F=$FIN
```

```
      SV2=$SV1/NN
      IF(L.GE.LL)GOTO  33
      IF(SV2.GE.EPS)  GOTO 13
   33 DT=2.*DT1
      RETURN
      END
```

Now we explain the second part of the subprogram which has the same structure as part 1 and part 3.

The statement PURGE $J, $JW, $FIN sets the state (not the value!) of the asynchronous variables $J, $JW and $FIN to "empty" regardless of their previous state. Next the loop counter $J and the process counter $JW are set to "full". In the serial version we start with J = 2, so here $J = 2, and because we did not initialize any process until now, we have $JW = 0.

```
      SUBROUTINE OPA($J,$JW,$FIN,NPROC1)
      PARAMETER (IZ=12)

      REAL BX(IZ,IZ),BY(IZ,IZ),VX(IZ,IZ),VY(IZ,IZ),
     *     P(IZ,IZ),DPX(IZ,IZ),DPY(IZ,IZ)
      REAL VB(IZ,IZ)
      REGISTER REAL DT3,DT4
      COMMON /HAIN/ BX,BY,VX,VY,P
      COMMON /CONI/ N,N1,N2,L,N11,NN.LL
      COMMON /CONR/ SDT,DT,DX2,DX,EPS,SV2
      COMMON /MHDP/ DPX,DPY,VB
      COMMON /MHD1/ DT1,DT2
      LOGICAL $FIN
      REGISTER INTEGER I,J
      DT3=DT1
      DT4=DT2
    1 J=$J
      $J=J+1
      IF(J.GT.N1)  GOTO 20
      DO     10       I=2,N1
      P(I,J)=P(I,J)*(1.-DT3*(VX(I,J)+VX(I,J-1)-VX(I-1,J)-VX(I-1,J-1)+
     *                VY(I,J)+VY(I-1,J)-VY(I,J-1)-VY(I-1,J-1)))
     *                -DT4*((VX(I,J)+VX(I-1,J)+VX(I,J-1)+VX(I-1,J-1))
     *                *(DPX(I,J)+DPX(I,J-1)+DPX(I-1,J)+DPX(I-1,J-1))
     *                +(VY(I,J)+VY(I-1,J)+VY(I,J-1)+VY(I-1,J-1))
     *                *(DPY(I,J)+DPY(I-1,J)+DPY(I,J-1)+DPY(I-1,J-1)))
      BX(I,J)=BX(I,J)+(VB(I,J)-VB(I,J-1))
      BY(I,J)=BY(I,J)-(VB(I,J)-VB(I-1,J))
   10 CONTINUE
      GOTO 1
   20 JW1=$JW+1
      IF(JW1.EQ.NPROC1)  $FIN=.TRUE.
      $JW=JW1
      RETURN
      END
```

We now create NPROC-2 processes J = 2, NPROC1 plus a process which is created by the main program itself, running simultaneously with the subroutines.

The processes OPA intended to run in parallel, are written as standard FORTRAN subroutines. Because of the immaturity of the compiler at benchmark time (April 1982), the user himself, at that time, had to store variables I, J, DT3 and DT4 into registers. The loop counter $J is emptied and increased for the next process. At the beginning J.LT.N1 is true, and the process under consideration starts the calculation of the pressure, and then returns to 1, empties, and increases the loop counter. All other processes do likewise, until there is one final process for which there is no more work to be done. For this process J.GT.N1 is true and the process counter $JW is emptied. Since $FIN is empty, the program then has to wait. Meanwhile the other processes are finishing their calculations and increasing the loop and the process counter until $JW1 has reached the number of initialized processes. Only this last process sets the state of the asynchronous variable $FIN to full, the statement F = $FIN can be executed and the program continues.

5. Restructuring of Basic Linear Algebraic Algorithms

Having obtained some useful information from the flow analysis options as described in section 3.1, and the vectorization of most of the DO loops of the central portion of the program has been accomplished, most of the work seems to have been done. Until now, though, we have not considered the algebraic structure of the basic algorithm. This is by far the most important consideration. Indeed, a good algorithm poorly coded is usually preferable to a poor one optimally coded |83|. For this reason, in the following chapters, we turn our attention to the restructuring of some basic linear algebraic algorithms such as matrix*vector and matrix*matrix, linear recursions up to more complex algorithms such as Gaussian elimination for full matrices, and iterative methods for the solution of linear algebraic systems of equations with sparse matrices (cf. |2|, |4|, |21|, |25|, |26|, |27|, |28|, |30|, |31|, |32|, |33|, |34|, |47|, |51|, |61|, |67|, |69|, |81|, |102|, |111|, |125|, |126|, |137|, |170|, |179|, |181|, |197|, |198| and |204| for sparse systems).

Problems involving linear algebraic calculations consume large quantities of computer time. If substantial improvements can be found for the linear algebraic section, a significant reduction in the overall computation time will be realized.

5.1 Basic Vector Operations

Many linear algebraic algorithms contain one or more of the common operations

vector * vector
matrix * vector
matrix * matrix.

The first one is easily vectorized by the current compilers but the latter two operations will cause some problems. In their original form the operational structure is more or less serial depending on the architecture of the computer. The matrix*vector procedure

$$\begin{pmatrix} Y_1 \\ Y_2 \\ \vdots \\ Y_N \end{pmatrix} = \begin{pmatrix} A_{11} & A_{12} & \cdots & A_{1N} \\ A_{21} & A_{22} & \cdots & A_{2N} \\ \vdots & \vdots & & \vdots \\ A_{N1} & A_{N2} & \cdots & A_{NN} \end{pmatrix} * \begin{pmatrix} X_1 \\ X_2 \\ \vdots \\ X_N \end{pmatrix}$$

results in

$$Y_i = \sum_{k=1}^{N} A_{ik} X_k \quad , \qquad i = 1,2,\ldots,N .$$

Here A_{ik} are the elements of an N * N - matrix A, and X, Y are vectors with N components. In other words the i-th component of Y is calculated by

$$Y_i = (\text{i-th row of A}) * X \quad ,$$

which for some vector computers (mainly memory-to-memory machines and a FORTRAN environment) is not suitable, since the elements are stored by columns. But looking at the operation more globally

$$\begin{aligned} Y_1 &= A_{11}X_1 + A_{12}X_2 + \cdots + A_{1N}X_N \\ Y_2 &= A_{21}X_1 + A_{22}X_2 + \cdots + A_{2N}X_N \\ &\vdots \\ Y_N &= A_{N1}X_1 + A_{N2}X_2 + \cdots + A_{NN}X_N \end{aligned}$$

yields the following basic vector structure

$$\begin{pmatrix} Y_1 \\ Y_2 \\ \vdots \\ Y_N \end{pmatrix} = \begin{pmatrix} A_{11} \\ A_{21} \\ \vdots \\ A_{N1} \end{pmatrix} *X_1 + \begin{pmatrix} A_{12} \\ A_{22} \\ \vdots \\ A_{N2} \end{pmatrix} *X_2 + \cdots + \begin{pmatrix} A_{1N} \\ A_{2N} \\ \vdots \\ A_{NN} \end{pmatrix} *X_N \quad .$$

The elements of each column of the matrix A are stored contiguously in the memory. The vectorized algorithm then has the form

$$\begin{pmatrix} Y_1 \\ Y_2 \\ \vdots \\ Y_N \end{pmatrix} = \begin{pmatrix} Y_1 \\ Y_2 \\ \vdots \\ Y_N \end{pmatrix} + X_j * \begin{pmatrix} A_{1j} \\ A_{2j} \\ \vdots \\ A_{Nj} \end{pmatrix} \qquad \text{for } j = 1,2, \ldots ,N .$$

In a previous step all elements of the vector Y are set to zero. The result is a triadic operation of the form

vector = vector + scalar * vector.

For many vector computers, the performance rate of these triads is almost doubled for long vectors (see e.g. Table 2.3).

For the matrix*vector case re-coding of the corresponding FORTRAN-code is very simple. One only has to interchange the loop-indices I and J to transform a rowlike algorithm into a columnlike one and obtain linked triads:

Scalar loop:

```
      DO   1   I = 1,N
      DO   1   J = 1,N
    1 Y(I) = Y(I) + X(J) * A(I,J)
```

Vector loop:

```
      DO   1   J = 1,N
      DO   1   I = 1,N
    1 Y(I) = Y(I) + X(J) * A(I,J)   .
```

The scalar inner loop represents a product of two vectors X and A(I,•) while the vector inner loop results in a triad for each J.

This principle is easily carried over to more complex problems as we shall shortly demonstrate using the matrix multiplication for large, full matrices. Matrix multiplication has three loops: an inner, middle, and outer. In the inner loop the arithmetic operations take place. By fixing the inner loop expression and varying the loop indices I,J, and K, six variants are possible for arranging the three loop indices. Each differs in how access is to be made to the matrix elements, i.e. by row or by column, as a scalar, vector, or matrix. Each permutation will have a different memory access pattern, which will have an important impact on its performance on a vector processor.

Given the N x N-matrices A and B, the formation of the product A * B = C in the usual manner is

```
      DO   1   I = 1,N
      DO   1   J = 1,N
      DO   1   K = 1,N
    1 C(I,J)   = C(I,J) + A(I,K) * B(K,J)
```

after having set C(I,J) = 0. for all I,J in a former step. The vector operation and data organization is described graphically by means of a diagram introduced by Dongarra |41|

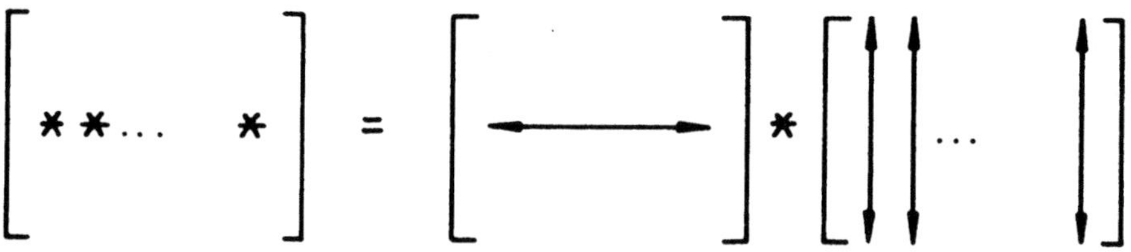

For the permutation of I and J, reference to the data is made slightly differently, so that the diagram is of the form

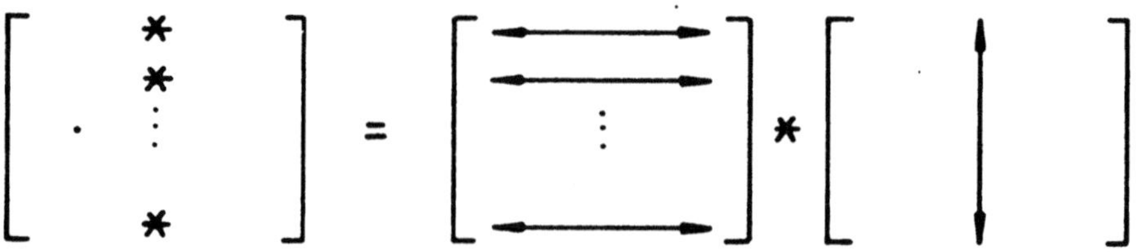

Both algorithms of the forms IJK and JIK are related by the fact the the inner loop is performing an inner product calculation. For reasons

concerning bank conflicts when access is made to elements in a row of a matrix, the use of inner products is not recommended on machines similar to the CRAY-1. On the other hand, since an inner product machine instruction exists on the memory-to-memory machine CYBER 205, one might be tempted to use it. But this does not result in the quickest procedure.

The algorithms of the form KIJ and KJI are related in that they take a multiple of a vector and add it to another vector as a basic operation. For the form KIJ the access pattern appears as:

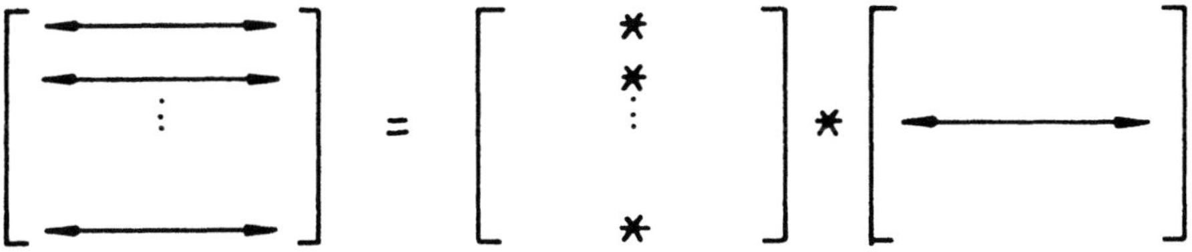

Every row of B is scaled by an element of A, and the result is used to update a row of C. For the form KJI we have:

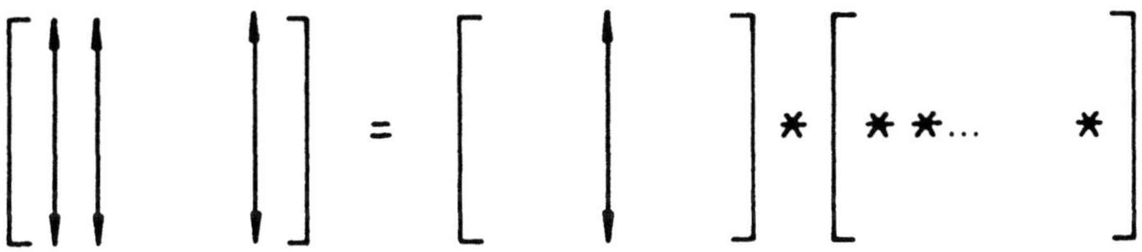

These algorithms are not to be recommended on vector computers since they up-date or load and store the entire matrix N times for each pass through the outer loop.

The final two possibilities are IKJ and JKI. Here the access patterns appear as:

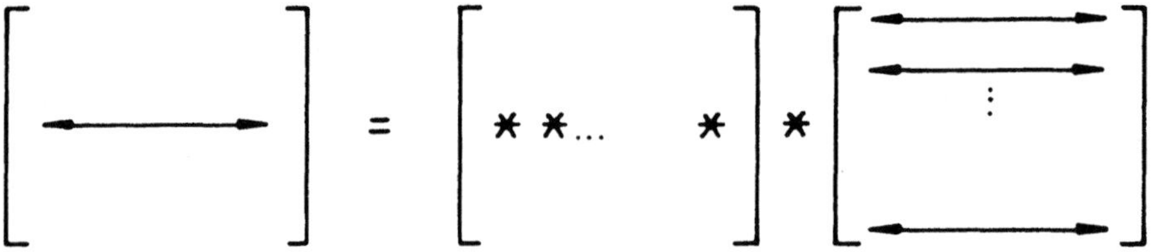

for IKJ and

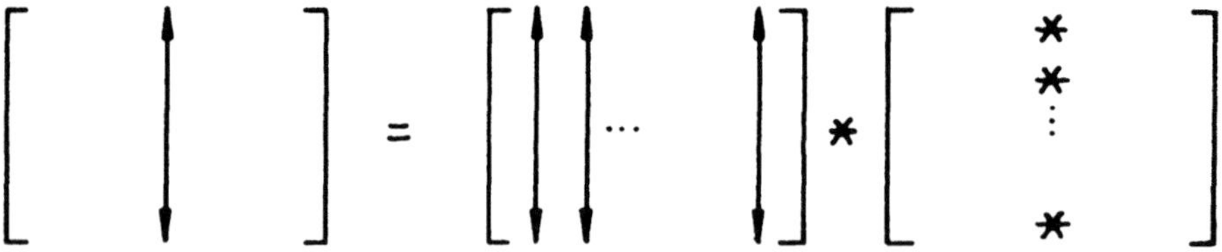

for JKI. These forms are very efficient on the current vector computers for full matrices, when N is sufficiently large. They use multiples of a set of vectors and accumulate the results in a single vector before storing the vector. Both forms are suitable for a "CRAY like" machine, while latter is superior for "CYBER like" machines owing to the columnwise operational scheme. As an illustration, we will discuss the latter form JKI in more detail.

Instead of looking only at one element C_{IJ} of the first form IJK, we write down all N results C_{IJ} for fixed J:

$$
\begin{aligned}
C_{1J} &= A_{11}B_{1J} + A_{12}B_{2J} + \ldots + A_{1N}B_{NJ} \\
C_{2J} &= A_{21}B_{1J} + A_{22}B_{2J} + \ldots + A_{2N}B_{NJ} \\
&\vdots \\
C_{NJ} &= A_{N1}B_{1J} + A_{N2}B_{2J} + \ldots + A_{NN}B_{NJ} \; .
\end{aligned}
$$

Again a basic column-wise vector structure is recognizable:

$$\begin{pmatrix} C_{1J} \\ C_{2J} \\ \vdots \\ C_{NJ} \end{pmatrix} = \begin{pmatrix} A_{11} \\ A_{21} \\ \vdots \\ A_{N1} \end{pmatrix} * B_{1J} + \dots + \begin{pmatrix} A_{1N} \\ A_{2N} \\ \vdots \\ A_{NN} \end{pmatrix} * B_{NJ}$$

for the J-th column of C. The vectorized algorithm then has the form

$$\begin{pmatrix} C_{1J} \\ C_{2J} \\ \vdots \\ C_{NJ} \end{pmatrix} = \begin{pmatrix} C_{1J} \\ C_{2J} \\ \vdots \\ C_{NJ} \end{pmatrix} + B_{KJ} * \begin{pmatrix} A_{1K} \\ A_{2K} \\ \vdots \\ A_{NK} \end{pmatrix} \qquad K = 1,2, \dots, N .$$

for every fixed J = 1,2...,N. For completeness, we add the following FORTRAN-version which corresponds to the above mentioned form JKI:

```
      DO  1  J = 1,N
      DO  1  K = 1,N
      DO  1  I = 1,N
    1 C(I,J)   = C(I,J) + A(I,K) * B(K,J)
```

B(K,J) is a constant for the innermost loop which again represents a triadic operation

vector = vector + vector * scalar.

5.2 Matrix Multiplication for Banded Matrices

The main disadvantage of the usual approaches to matrix multiplication, discussed in the last section, is that they become inefficient for banded matrices with relatively narrow bandwidths. We will demonstrate this fact by means of the matrix multiplication of sparse three-diagonal matrices and the application of the form JKI which results in a pentadiagonal matrix with the column vectors

$$\underline{C}_K = A_{K-1,K} \cdot \underline{B}_{K-1} + A_{KK} \cdot \underline{B}_K + A_{K+1,K} \cdot \underline{B}_{K+1}$$

with

$$\underline{C}_K = \begin{pmatrix} C_{K-2,K} \\ C_{K-1,K} \\ C_{KK} \\ C_{K+1,K} \\ C_{K+2,K} \end{pmatrix}$$

and

$$\underline{B}_{K-1} = \begin{pmatrix} B_{K-2,K-1} \\ B_{K-1,K-1} \\ K_{K,K-1} \\ 0 \\ 0 \end{pmatrix} \qquad \underline{B}_K = \begin{pmatrix} 0 \\ B_{K-1,K} \\ B_{KK} \\ B_{K+1,K} \\ 0 \end{pmatrix} \qquad \underline{B}_{K+1} = \begin{pmatrix} 0 \\ 0 \\ B_{K,K+1} \\ B_{K+1,K+1} \\ B_{K+2,K+1} \end{pmatrix} .$$

Although the vector operations to evaluate the vector C_K can be treated as a linked triad for every K, namely

$$\underline{C}_K = \underline{C}_K + A_{IK} \underline{B}_I \quad , \qquad I = K-1,K,K+1 \quad ,$$

the gain in performance is poor, resulting from the maximum vector length of 5.

The key to the solution of this problem is to store the matrices by diagonals instead of by rows or columns. This storing seems natural for large banded matrices in the sense that the matrix is defined and stored in terms of very few vectors which are as long as possible. A second advantage is that the transposed matrix A^T of a matrix A is readily available in terms of the same diagonal vectors whose elements are stored consecutively in the memory.

The basic idea is that instead of forming a column of C, we will form a diagonal of C. Let us consider again the multiplication of tridiagonal matrices. The result is a matrix with five diagonals

$$\underline{c}_{-2}\ ,\ \underline{c}_{-1}\ ,\ \underline{c}_{0}\ ,\ \underline{c}_{1}\ ,\ \underline{c}_{2}$$

where

$$\underline{c}_{-2} = \begin{pmatrix} 0 \\ 0 \\ c_{31} \\ \vdots \\ c_{K,K-2} \\ \vdots \\ c_{N,N-2} \end{pmatrix}\ ,\ \underline{c}_{-1} = \begin{pmatrix} 0 \\ c_{21} \\ \vdots \\ c_{K,K-1} \\ \vdots \\ c_{N,N-1} \end{pmatrix}\ ,\ \underline{c}_{0} = \begin{pmatrix} c_{11} \\ \vdots \\ c_{KK} \\ \vdots \\ c_{NN} \end{pmatrix}$$

$$\underline{c}_{1} = \begin{pmatrix} c_{12} \\ \vdots \\ c_{K,K+1} \\ \vdots \\ c_{N-1,N} \\ 0 \end{pmatrix}\ ,\ \underline{c}_{2} = \begin{pmatrix} c_{13} \\ \vdots \\ c_{K,K+2} \\ \vdots \\ c_{N-2,N} \\ 0 \\ 0 \end{pmatrix} .$$

The zeros in the vectors may be omitted but are used here to obtain a homogeneous structure. The diagonals of A are defined as follows

$$\underline{A}_{-1} = \begin{pmatrix} 0 \\ A_{21} \\ \vdots \\ A_{K,K-1} \\ \vdots \\ A_{N,N-1} \\ 0 \end{pmatrix}, \quad \underline{A}_0 = \begin{pmatrix} A_{11} \\ \vdots \\ A_{KK} \\ \vdots \\ A_{NN} \\ 0 \end{pmatrix}, \quad \underline{A}_1 = \begin{pmatrix} 0 \\ A_{12} \\ \vdots \\ A_{K,K+1} \\ \vdots \\ A_{N-1,N} \\ 0 \end{pmatrix}$$

and analogously for B. Using the nomenclature of |132|, V(p;q) will denote the vector

$$V(p;q) = \begin{pmatrix} V_{p+1} \\ \vdots \\ V_{M-q} \end{pmatrix}$$

where M is the vector length of V. We say that V(p;q) has been obtained by displacing the vector V by p from the top and q from the bottom. The following result can be verified in a straightforward computation

$$\underline{C}_{-2} = \underline{A}_{-1}(2;1) \cdot \underline{B}_{-1}(1;2)$$

$$\underline{C}_{-1} = \underline{A}_{-1}(1;1) \cdot \underline{B}_0(0;2) + \underline{A}_0(1;1) \cdot \underline{B}_{-1}(1;1)$$

$$\underline{C}_0 = \underline{A}_{-1}(1;0) \cdot \underline{B}_1(1;0) + \underline{A}_0(0;1) \cdot \underline{B}_0(0;1) + \underline{A}_1(1;0) \cdot \underline{B}_{-1}(1;0)$$

$$\underline{C}_1 = \underline{A}_0(0;2) \cdot \underline{B}_1(1;1) + \underline{A}_1(1;1) \cdot \underline{B}_0(1;1)$$

$$\underline{C}_2 = \underline{A}_1(1;2) \cdot \underline{B}_1(2;1) \quad .$$

The algorithm is not restricted to the multiplication of tridiagonal matrices. Analogous algorithms for more general banded matrices may be found in |132|. There a "FORTRAN-like" algorithm can be found and its limits discussed. For full matrices, the start-up costs are about three times higher for the diagonal algorithm compared to the conventional algorithms described above. But for narrow banded matrices the diagonal algorithm is much more efficient. For the particular case of the multiplication of two tridiagonal matrices with N = 1000 on the STAR-100 the diagonal algorithm is about 48.5 times faster than the conventional one. For pentadiagonal matrices this factor decreases to 33.2. The cross-over point is found to be at a bandwidth of r = 0.5N.

5.3 Gaussian Elimination for the Solution of Algebraic Systems with Full Matrices

Linear, non-homogeneous, algebraic systems of equations with full matrices arise frequently in numerical analysis,e.g. in panel methods (see |230|) and in Galerkin methods (see section 9.4). Numerical methods for solving these problems are divided into two classes: exact and iterative.

In this section we discuss the vectorization of an exact method, Gaussian elimination without pivoting, which gives a solution of the problem by using a finite number of elementary arithmetic operations. Vectorization of the method with relative pivoting is discussed in |168|. The algorithm consists of a series of successive eliminations by means of which the given system is transformed into a system with a triangular matrix whose solution presents no further difficulty.

Consider the system of equations given by:

$$
\begin{aligned}
a_{11}x_1 + a_{12}x_2 + \dots + a_{1n}x_n &= a_{1,n+1} \\
a_{21}x_1 + a_{22}x_2 + \dots + a_{2n}x_n &= a_{2,n+1} \\
&\vdots \\
a_{n1}x_1 + a_{n2}x_2 + \dots + a_{nn}x_n &= a_{n,n+1} .
\end{aligned}
$$

Assume that the matrix of the system is non-singular and $a_{ii} \neq 0$. The elimination of the unknown x_1 in rows no. 2, 3, ..., n is performed by adding a multiple of the first row to the corresponding row including the right hand side, in detail:

$$
i\text{ - th row} = 1\text{-th row} * \left(-\frac{a_{i1}}{a_{11}}\right) + i\text{-th row} .
$$

This classical computational scheme has a rowwise structure and is therefore not appropriate for many vector computers. However, looking more globally at this scheme, we obtain:

$$\text{2-th row} = \text{1-th row} * \left(-\frac{a_{21}}{a_{11}}\right) + \text{2-th row}$$

$$\text{3-th row} = \text{1-th row} * \left(-\frac{a_{31}}{a_{11}}\right) + \text{3-th row}$$

$$\vdots$$

$$\text{n-th row} = \text{1-th row} * \left(-\frac{a_{n1}}{a_{11}}\right) + \text{n-th row} .$$

For every first element a_{i1} of the rows i = 2,3,...,n, this becomes:

$$a_{i1}^{(1)} = a_{11} * \left(-\frac{a_{i1}}{a_{11}}\right) + a_{i1} = 0 .$$

The index (1) means first elimination step. For every second element a_{i2} of the rows i = 2,3,...,n, we obtain the elements:

$$a_{i2}^{(1)} = a_{12} * \left(-\frac{a_{i1}}{a_{11}}\right) + a_{i2}$$

or explicitly:

$$a_{22}^{(1)} = - a_{21} * \left(\frac{a_{12}}{a_{11}}\right) + a_{22}$$

$$a_{32}^{(1)} = - a_{31} * \left(\frac{a_{12}}{a_{11}}\right) + a_{32}$$

$$\vdots$$

$$a_{n2}^{(1)} = - a_{n1} * \left(\frac{a_{12}}{a_{11}}\right) + a_{n2}$$

and for every third element analogously:

$$a_{23}^{(1)} = - a_{21} * \left(\frac{a_{13}}{a_{11}}\right) + a_{23}$$

$$a_{33}^{(1)} = - a_{31} * \left(\frac{a_{13}}{a_{11}}\right) + a_{33}$$

$$\vdots$$

$$a_{n3}^{(1)} = -a_{n1} * \left(\frac{a_{13}}{a_{11}}\right) + a_{n3}$$

and so on. For every K-th element a_{iK} of the rows $i = 2,3,\dots,n$, we arrive at the general formula:

$$\begin{pmatrix} a_{2K}^{(1)} \\ a_{3K}^{(1)} \\ \vdots \\ a_{nK}^{(1)} \end{pmatrix} = - \begin{pmatrix} a_{21} \\ a_{31} \\ \vdots \\ a_{n1} \end{pmatrix} * \frac{a_{1K}}{a_{11}} + \begin{pmatrix} a_{2K} \\ a_{3K} \\ \vdots \\ a_{nK} \end{pmatrix},$$

which is a linked triad of the form:

vector * scalar + vector

for all $K = 2,3,\dots,n+1$. Thus, we have transformed the first elimination step into a columnar algorithm. The augmented matrix now has the following form:

$$\begin{pmatrix} a_{11} & a_{12} & a_{13} & \cdots\cdots & a_{1,n+1} \\ 0 & a_{22}^{(1)} & a_{23}^{(1)} & \cdots\cdots & a_{2,n+1}^{(1)} \\ \vdots & & & & \\ 0 & a_{n2}^{(1)} & a_{n3}^{(1)} & \cdots\cdots & a_{n,n+1}^{(1)} \end{pmatrix}.$$

The second step of the elimination of the second unknown x_2 in all the rows $i = 3,4,\dots,n$ proceeds similarly to obtain the linked triads:

$$\begin{pmatrix} a_{3K}^{(2)} \\ a_{4K}^{(2)} \\ \vdots \\ a_{nK}^{(2)} \end{pmatrix} = - \begin{pmatrix} a_{32}^{(1)} \\ a_{42}^{(1)} \\ \vdots \\ a_{n2}^{(1)} \end{pmatrix} * \frac{a_{2K}^{(1)}}{a_{22}^{(1)}} + \begin{pmatrix} a_{3K}^{(1)} \\ a_{4K}^{(1)} \\ \vdots \\ a_{nK}^{(1)} \end{pmatrix}$$

for all $K = 3,4,\dots,n+1$. One proceeds in this manner until, at the last step, we obtain the equations:

$$\left(a_{nK}^{(n-1)}\right) = -\left(a_{n,n-1}^{(n-2)}\right) * \frac{a_{n-1,K}^{(n-2)}}{a_{n-1,n-1}^{(n-2)}} + \left(a_{nK}^{(n-2)}\right)$$

for K = n and n+1. In FORTRAN, this vectorized elimination step can be coded as follows

```
      DO    1   L = 2,N
      AL   = - 1./ A(L-1,L-1)
      DO    1   K = L,NP1
      ALK =  A(L-1,K) * AL
      DO    1       I = L,N
    1 A(I,K)= A(I,L-1) * ALK + A(I,K)
```

where NP1 = N + 1. The innermost DO-loop consists of a linked triad of the form:

vector * scalar + vector.

Combining all the first equations of each step results in the system with triangular matrix:

$$
\begin{aligned}
a_{11}x_1 + a_{12}x_2 + a_{13}x_3 + \dots + a_{1n}x_n &= a_{1,n+1} \\
a_{22}^{(1)}x_2 + a_{23}^{(1)}x_3 + \dots + a_{2n}^{(1)}x_n &= a_{2,n+1}^{(1)} \\
a_{33}^{(2)}x_3 + \dots + a_{3n}^{(2)}x_n &= a_{3,n+1}^{(2)} \\
&\vdots \\
a_{nn}^{(n-1)}x_n &= a_{n,n+1}^{(n-1)} .
\end{aligned}
$$

This system is equivalent to the original system. Hence, to solve a given system using the elimination scheme, one first constructs an auxiliary triangular system of equations, which is then solved straightforward.

From the triangular system we find the values for the unknowns in sequence from x_n to x_1: Knowing the value of x_n from the last equation, x_{n-1} can be evaluated from the (n-1)th equation, x_{n-2} can be evaluated from the (n-2)th equation, and so on. The backward recursive formula for obtaining the values of the solution x_i is:

$$x_i = \left(a_{i,n+1}^{(i-1)} - \sum_{K=i+1}^{n} a_{iK}^{(i-1)} x_K\right) \Big/ a_{ii}^{(i-1)} .$$

In this form, the back-substitution is recursive. Again, a possible vectorization is achieved by looking at the algorithm as a whole. In the i-th step, the row-wise division by $a_{ii}^{(i-1)}$ is first performed. Omitting the upper indices, the back-substitution then becomes:

$$x_n = a_{n,n+1}$$

$$x_{n-1} = a_{n-1,n+1} - a_{n-1,n} x_n$$

$$x_{n-2} = a_{n-2,n+1} - a_{n-2,n} x_n - a_{n-2,n-1} x_{n-1}$$

$$\vdots$$

$$x_1 = a_{1,n+1} - a_{1,n} x_n - a_{1,n-1} x_{n-1} - \dots - a_{1,2} x_2 \quad .$$

Now the x_i can be evaluated simultaneously. A first step is

$$x_n = a_{n,n+1}$$

$$x_{n-1} = a_{n-1,n+1}$$

$$\vdots$$

$$x_1 = a_{1,n+1}$$

leading the exact value of x_n, which is then substituted into the equations for $x_{n-1}, x_{n-2}, \dots, x_1$, so that we have at the second step:

$$x_{n-1} = x_{n-1} - a_{n-1,n} * x_n$$

$$x_{n-2} = x_{n-2} - a_{n-2,n} * x_n$$

$$\vdots$$

$$x_1 = x_1 - a_{1,n} * x_n \quad .$$

Here the value of x_{n-2} is exact. Continuing we arrive finally at the equation:

$$x_1 = x_1 - a_{12} * x_2 \quad .$$

This columnar procedure easily can be translated into a vectorized FORTRAN as follows

```
  DO    1    I = 1,N
1 X(I) =  A(I,N+1)
  DO    2    K = 2,N
  NK    =  N-K+1
  DO    2    I = 1,NK
2 X(I) = X(I) - A(I,NK+1) * X(NK+1).
```

Again the innermost DO-loop consists of a linked triad of the form

vector + vector * scalar.

It is often useful to check the accuracy of the solution vector x, above all for large systems of equations (say $n > 100$), by any norm $||\cdot||$ of the error vector $\varepsilon = x^* - x$ with the exact solution x^*. However, the error vector may not be computable without knowledge of the exact solution x^*. A vector characterizing the exactness of the approximate solution x of the system is the residual vector:

$$r = b - Ax \quad .$$

It is clear that

$$r = b - Ax = b - A(x^* - \varepsilon) = A\varepsilon \, .$$

Thus relaxation may be set up to achieve $||r|| \rightarrow 0$.

Let $x^{(0)} = x$ be the approximate solution of a system $Ax^* = b$ computed using Gaussian elimination. Suppose we can improve the approximate solution by adding a "small" vector $z^{(1)}$ with:

$$x^* = x^{(0)} + z^{(1)}$$

and x^* then is the exact solution of the system. We get the equation:

$$Az^{(1)} = A(x^* - x^{(0)}) = b - Ax^{(0)} = r^{(0)}$$

which we have to solve for $z^{(1)}$. The elimination step, however, has been performed previously, so that we only need to calculate the transformed right-hand side and perform the back-substitution, both of which are easily vectorized. For the transformation of the right-hand side, the multiples $a_{ik}^{(i-1)}/a_{ii}^{(i-1)}$ must be stored during the previous elimination step. The back-substitution then proceeds as described above.

Another application of the Gaussian elimination method is the calculation of the inverse of a non-singular matrix A. Let the columns of the inverse of A be denoted by $b_1, b_2, \ldots, b_n$, and the columns of the identity matrix be denoted by $e_1, e_2, \ldots, e_n$. By the definition of the inverse we have:

$$A * A^{-1} = I$$

$$A * (b_1, b_2, \ldots, b_n) = (e_1, e_2, \ldots, e_n)$$

which is equivalent to the solution of the n systems:

$$Ab_1 = e_1$$

$$Ab_2 = e_2$$

$$\vdots$$

$$Ab_n = e_n \, .$$

These systems have the same matrix of coefficients, so that Gaussian elimination can be applied simultaneously using n right-hand side vectors. The augmented matrix now has 2n columns instead of n + 1 when solving only one system.

5.4 Linear and Nonlinear Recurrences

The vectorization of algorithms including linear and nonlinear recurrences such as Thomas-, Richtmyer- and Rusanov-algorithm (see e.g. |67|, |225|, and section 9.2) for the solution of large, linear systems of algebraic equations with tridiagonal or block tridiagonal matrices is in general not straightforward. However, to avoid interdependencies of neighbouring grid points in two- and three-dimensional problems, a so-called Red-Black or ZEBRA-pattern in one dimension (cf. next chapter) and a parallel evaluation of the recurrences in the other dimension(s) solve the problem in a rather elementary way with only a few modifications to an already existing program, so allowing vector operations on long and contiguously stored vectors and arrays, and maintaining not only the efficiency of the corresponding recursive algorithm, but also the high potential of the vector machine (cf. |67|, |69|, |113|, |114|, |115|, |117|, |142| and |173|).

In the following we shall solve systems of n algebraic equations Au = f with the tridiagonal matrix:

$$(5.1) \qquad A = \begin{pmatrix} a_1 & c_1 & & & \\ b_2 & a_2 & c_2 & & \\ & \ddots & \ddots & \ddots & \\ & & b_{n-1} & a_{n-1} & c_{n-1} \\ & & & b_n & a_n \end{pmatrix}$$

which often arise in practice, e.g. when solving ordinary or partial differential equations with second order derivatives by discrete numerical methods. There are a number of related methods for solving this system serially in a time proportional to n. One of these methods is Gaussian elimination, which for tridiagonal systems reduces to the so-called Thomas-algorithm. It is very efficient with 5n - 4 arithmetic operations compared to $n(n^2+3n-1)/3$ operations for the complete Gaussian elimination. We explicitly assume, that the LU decomposition of A into the product of a lower triangular matrix L and an upper triangular matrix U exists. That is, A = L * U where:

$$(5.2)\qquad L = \begin{pmatrix} 1 & & & \\ \gamma_2 & 1 & & \\ & \ddots & \ddots & \\ & & \gamma_n & 1 \end{pmatrix}, \qquad U = \begin{pmatrix} \alpha_1 & c_1 & & & \\ & \alpha_2 & c_2 & & \\ & & \ddots & \ddots & \\ & & & \alpha_{n-1} & c_{n-1} \\ & & & & \alpha_n \end{pmatrix}.$$

After computing L and U, it is relatively straightforward to solve the resulting triangular systems of equations:

$$(5.3)\qquad Ly = f \quad , \quad Uu = y.$$

The whole algorithm can be expressed in three stages:

decomposition:

$$\alpha_1 = a_1$$

$$\gamma_i = b_i/\alpha_{i-1} \qquad i = 2,3,\ldots,n$$

$$\alpha_i = a_i - c_{i-1}\gamma_i$$

(5.4) forward substitution:

$$g_1 = f_1$$

$$g_i = f_i - \gamma_i g_{i-1} \qquad i = 2,3,\ldots,n$$

backward substitution:

$$u_n = g_n/\alpha_n$$

$$u_i = (g_i - c_i u_{i+1})/\alpha_i \qquad i = n-1,\ldots,1$$

The algorithm is stable, if

$$|a_1| > |c_1| > 0$$

$$|a_i| \geq |b_i| + |c_i| \;,$$

$$b_i \cdot c_i \neq 0$$

$$|a_n| > |b_n| > 0 \;,$$

which in many applications can be fulfilled. If one of the conditions:

$$c_1 \neq 0\,, \quad b_i c_i \neq 0\;, \quad b_n \neq 0$$

is violated, the system can be reduced to two or more smaller systems which are essentially uncoupled.

In solving systems of m partial differential equations there arise systems of algebraic equations of the form (5.1), but with a block tridiagonal matrix A where the a_i, b_i, c_i are now m x m -matrices. The numerical treatment of the boundary-layer equations for two dimensional, incompressible, viscous flow leads to systems with 2x2-blocks, whereas 5x5-blocks arise in the solution of the Navier-Stokes equations for three-dimensional compressible viscous flows. In these cases the above mentioned Thomas-algorithm extends easily to the Richtmyer algorithm. The vectorization of the Richtmyer algorithm will be explained in section 9.2 in further detail.

Unfortunately the three loops in (5.4) are all sequential recurrences that must be evaluated one term at a time. The fact that the previous element must be known before the present one is computed, prevents the algorithm from taking any advantage of the vector hardware features on a computer since all elements of a vector must be known before a vector operation is initiated. Hence the algorithm (5.4),which is the fastest one solving tridiagonal systems on a serial computer, is highly unsuitable on a vector computer.

However in all cases that occur in applications there are alternatives repairing these problems. In the following we point out five remedies (besides those mentioned in chapter 3 which seem to be among the most important ones:

a) restructuring of the existing serial algorithm
b) solving a set of tridiagonal systems simultaneously
c) iteration of the recurrences due to Traub |204|
d) evaluating the linear first-order recurrences by the method of cyclic reduction, and the non-linear recursion in (5.4) by the method of recursive doubling
e) use of other algorithms

a) In general solving k tridiagonal systems on a vector computer, one may apply the best sequential algorithm to the k systems simultaneously or take a good vector or parallel algorithm for solving one single system after the other. The first way will be described in part b) and the second will be explained in part d). Here we show how to modify the existing sequential algorithm with few modifications, to introduce a higher level of parallelism into the linear and non-linear recurrences as they occur in (5.4). As an example we look at the first highly non-linear recurrence. With $d_i = 1/\alpha_i$ we reduce the number of time consuming divisions. Combining d_i and γ_i of (5.4) we get:

$$(5.5)\qquad \begin{aligned} d_i &= 1/(a_i - b_i c_{i-1} d_{i-1}) \\ \gamma_i &= b_i d_{i-1} \end{aligned} \qquad i = 2,3,\ldots,n\ .$$

The quantities $b_i c_{i-1}$ could be computed before the evaluation of the d_i begins. Also computation of the γ_i could be delayed until all the values d_i have been obtained. Thus both of these calculations are vector multiplication operations:

$$(5.6)\qquad \begin{aligned} e_i &= b_i c_{i-1} && i = 2,3,\ldots,n \\ d_1 &= 1/a_1 \\ d_i &= 1/(a_i - e_i d_{i-1}) && i = 2,3,\ldots,n \\ \gamma_i &= b_i d_{i-1} && i = 2,3,\ldots,n\ , \end{aligned}$$

Version (5.6) proposed by Lambiotte et al. |125| is faster than (5.5) on some vector computers for all $n>n_0$ (where n_0 depends on the computer), because of the start-up time in vector operations. For example $n_0 = 12$ on a

STAR-100. For vector computers with only short start-up time, however, no improvement will be obtained (e.g. CRAY-1S).

The third recurrence in (5.4) could be changed as in (5.9), so that the following algorithm for the solution of tridiagonal systems is more suitable on vector computers for $n>n_0$:

decomposition:

$$e_i = b_i c_{i-1} \qquad i = 2,3,\dots,n$$

$$d_1 = 1/a_1$$

(5.7)

$$d_i = 1/(a_i - e_i d_{i-1}) \qquad i = 2,3,\dots,n$$

$$\gamma_i = b_i d_{i-1} \qquad i = 2,3,\dots,n$$

forward substitution:

$$g_1 = f_1$$

(5.8)

$$g_i = f_i - \gamma_i g_{i-1} \qquad i = 2,3,\dots,n$$

backward substitution:

$$g_i = g_i \cdot d_i \qquad i = 1,2,\dots,n$$

$$h_i = c_i \cdot d_i \qquad i = 1,2,\dots,n-1$$

(5.9)

$$u_n = g_n$$

$$u_i = g_i - h_i u_{i+1} \qquad i = n-1,\dots,1 .$$

This example typifies one major disadvantage of vectorized codes, namely the increase in storage capacity required for temporary arrays such as here e_i and h_i.

b) If one has to solve a set of k tridiagonal systems, as frequently occurs in the solution of differential equations in more than one dimension, the Thomas-algorithm will be the best on a vector computer. If the k systems are independent, which is the case in all time-splitting and ADI-type difference methods (with implicit character in only one direction at each stage), one can change all variables in the algorithm into vectors of length k running across all tridiagonal systems.

In the case of difference methods which are implicit in more than one direction (joining neighbouring points in two or more dimensions such as the five point star for the Laplace operator, as described in the next chapter), one first has to implement a ZEBRA (odd-even line) pattern and to solve simultaneously the "black" systems and afterwards all the "white" systems, and so on. In so doing, the effectiveness of the algorithm is preserved and maximum vector performance is obtained (see chapter 6).

Unfortunately the total required storage increases almost k-fold compared with solving the tridiagonal systems one at a time. We therefore have to look for other methods suitable for vector computers if sufficient storage is not available.

c) The next step is a vectorization of the remaining serial loops in (5.7) to (5.9). One idea, (cf. Traub |204|), is to evaluate these recurrences iteratively. We shall demonstrate this principle using the following linear first-order recurrence

(5.10)
$$x_0 = 0$$
$$x_i = a_i + b_i \cdot x_{i-1} \qquad i = 1,2,\dots,n\,.$$

For the solution on a vector computer, we convert the sequential form into an iteration by successively approximating the x_i:

(5.11)
$$x_i^{(0)} = 0 \qquad i = 1,2,\dots,n$$
$$x_i^{(j)} = a_i + b_i x_{i-1}^{(j-1)} \qquad i = 1,2,\dots,n$$

for all iteration steps $j = 1,2,\dots,n$. Since all x_i are updated simultaneously, this can be viewed as a vector iteration and leads to the exact solution after exactly n iterations. Instead of 2n operations for the evaluation of the sequential recurrence (5.10), however, we get 2n(n-1) operations for (5.11), so that (5.11) requires many more operations than (5.10). However it becomes interesting in the context of a computer which can perform operations on vectors of length n much faster than it can perform n scalar operations.

It is obvious from (5.11) that $x_i^{(j)}$ depends only on $x_{i-1}^{(j-1)}$ and on no other component, and, more generally, the odd-subscripted variables depend only on the even-subscripted ones and vice-versa. If we re-define all the even-subscripted x_i and then all the odd-subscripted ones, always using the most recent values available, the new iteration will be almost twice as fast.

Beyond that, Heller et al. |81| proved that the iteration procedure (5.11) converges much faster than in n cycles to a certain accuracy if the tridiagonal matrix is strictly diagonally dominant (cf. |224|). For the solution of the tridiagonal systems arising in the numerical treatment of Poisson's equation (for n = 32) the evaluation of the recurrences to an accuracy of 10^{-10} was complete after 5 iterations for (5.7) and after 10 iterations for (5.8) and (5.9). A complete, detailed study of convergence may be found in |81|.

d) Another vectorization procedure is to replace the whole recurrence under consideration (not the complete algorithm!) by other algorithms which are more compatible with the computer hardware. Two approved methods in this case are cyclic reduction and recursive doubling for the solution of linear first order recurrences of the form (5.10).

Cyclic (odd-even) reduction (cf. |80|, |160|) is widly applied in numerical annalysis; particularly when one wishes to introduce parallelism into a problem. The above recurrence relates neighbouring terms in the sequence. The basic idea of cyclic reduction will be described in more detail in the next section in conection with the direct solution for tridiagonal systems. Here we use it to combine adjacent terms of the recurrence to relate x_i to x_{i-2}:

(5.12) $$x_i = a_i + b_i (a_{i-1} + b_{i-1} x_{i-2}) ,$$

which is also a linear, first-order recurrence of the form:

$$x_i = a_i^{(1)} + b_i^{(1)} x_{i-2}$$

with a new set of coefficients:

$$a_i^{(1)} = b_i a_{i-1} + a_i \qquad b_i^{(1)} = b_i b_{i-1} .$$

This process can be repeated to obtain recurrences relating every fourth term, every eight term, and so on. The result may be summarized by the reduced equations for each level ℓ running from 0 to $\log_2 n$:

(5.13) $$x_i = a_i^{(\ell)} + b_i^{(\ell)} x_{i-2^\ell}$$

where:

(5.14) $$a_i^{(\ell)} = b_i^{(\ell-1)} a_{i-2^{(\ell-1)}}^{(\ell-1)} + a_i^{(\ell-1)}$$
$$b_i^{(\ell)} = b_i^{(\ell-1)} b_{i-2^{(\ell-1)}}^{(\ell-1)}$$

with:

$$a_i^{(0)} = a_i \quad , \quad b_i^{(0)} = b_i \quad .$$

If the subscript of any x_i, a_i or b_i is outside the intervall $|1,n|$, its value is set to zero. When $\ell = \log_2 n$, the solution of the recurrence is:

(5.15) $$x_i = a_i^{(\log_2 n)} \quad .$$

We therefore have to generate the coefficients $a_i^{(\ell)}$ and $b_i^{(\ell)}$ successively, until $a_i^{(\log_2 n)}$ is found. The above equations can easily be evaluated in parallel. The values on the right-hand side are all known values computed at the previous level $(\ell-1)$.

In the remainder of this section we assume that n is a power of 2, but this is not an essential assumption. The idea of recursive doubling (cf. |43|,|197|) then is to rewrite the recurrence formula (5.10) so that x_{2i} is a function of x_i. Thus in successive iterations we can compute x_1, x_2, x_4, x_8, etc., and x_n can be computed in $\log_2 n$ iterations.

To begin, we substitute for x_{i-1} in (5.10) and find

$$x_1 = a_1 + b_1 x_0 = a_1$$

$$x_2 = a_2 + b_2 x_1 = a_2 + a_1 b_2$$

$$x_3 = a_3 + b_3 x_2 = a_3 + a_2 b_3 + a_1 b_2 b_3$$

$$\vdots$$

$$x_i = \sum_{j=1}^{i} a_j \prod_{K=j+1}^{i} b_K$$

where $\prod_{K=i+1}^{i} b_K = 1$.

Suppose that all the b_i are set equal to one. In this case x_i is the sum of the first i components $a_1,\ldots,a_i$ which is expressed as $x_i\ (a_1,\ a_2,\ldots,a_i)$. $x_{2i} = x_{2i}(a_1,a_2,\ldots,a_{2i})$ is defined similarly. Now we can express x_{2i} in terms of two functions that are each half as complex as $x_{2i}(a_1,a_2,\ldots,a_{2i})$, this being the central property of recursive doubling:

$$x_{2i}(a_1,\ldots,a_{2i}) = x_i(a_1,\ldots,a_i) + x_i(a_{i+1},\ldots,a_{2i})$$

where $x_i(a_{i+1},\ldots,a_{2i})$ is defined to be the sum of $a_{i+1},\ldots,\ a_{2i}$. Then we can evaluate both terms independently. They are computationally identical, and differ only in the values of their arguments. The iterative application of this procedure then results in the recursive doubling algorithm, which we demonstrate below with $n = 8$ and $b_i = 1$. First, given the column vector $A = (a_1,a_2,\ldots,a_8)^T$ (T indicating transpose), define a column vector so that $X = (x_1,x_2,\ldots,x_8)^T$. The next step is to use the iterated addition algorithm, whereby the i-th shifted vector corresponding to A is added to X:

$$\begin{aligned}
x_1 &= a_1\\
x_2 &= a_2+a_1\\
x_3 &= a_3+a_2+a_1\\
x_4 &= a_4+a_3+a_2+a_1\\
x_5 &= a_5+a_4+a_3+a_2+a_1\\
x_6 &= a_6+a_5+a_4+a_3+a_2+a_1\\
x_7 &= a_7+a_6+a_5+a_4+a_3+a_2+a_1\\
x_8 &= a_8+a_7+a_6+a_5+a_4+a_3+a_2+a_1.
\end{aligned}$$

After the first step (which was storing of A into X) the first element x_1 is correct. Then starting with the vector $X_7(8)$ with components $a_2,\ldots,a_8$ we add $X_7(8)$ to $X_7(7)$. Here the subscript I of X_I (J) means I consecutive elements of the vector X the highest component being the Jth component. After this second step, the result is:

$$X_7(8) = X_7(8) + X_7(7)$$

or explicitly:

$$\begin{aligned}
x_1 &= a_1 \\
x_2 &= a_2 + a_1 \\
x_3 &= a_3 + a_2 \\
x_4 &= a_4 + a_3 \\
x_5 &= a_5 + a_4 \\
x_6 &= a_6 + a_5 \\
x_7 &= a_7 + a_6 \\
x_8 &= a_8 + a_7 \quad .
\end{aligned}$$

The first two components of X are already exact. Looking again at the final result, the next step is simply to add the elements $x_1,\ldots,\ x_6$ to the elements $x_3,\ldots,x_8$ which results in:

$$X_6(8) = X_6(8) + X_6(6)$$

or

$$\begin{aligned}
x_1 &= a_1 \\
x_2 &= a_2 + a_1 \\
x_3 &= a_3 + a_2 + a_1 \\
x_4 &= a_4 + a_3 + a_2 + a_1 \\
x_5 &= a_5 + a_4 + a_3 + a_2 \\
x_6 &= a_6 + a_5 + a_4 + a_3 \\
x_7 &= a_7 + a_6 + a_5 + a_4 \\
x_8 &= a_8 + a_7 + a_6 + a_5 .
\end{aligned}$$

Here the elements $x_1,\ldots,x_4$ are exact. The last step is now straightforward. We have to add the first four elements of X to the last four ones and get the final result:

$X_4(8) = X_4(8) + X_4(4).$

A FORTRAN-like algorithm for the solution of the special recurrence (5.10) with b_i=1 would have the following simple form:

```
      DO    1  J = 1,N
    1 X(J)  =A(J)

      DO    2  I = 1,N2
      K = 2**(I-1)+1
      DO    2  J = K,N
    2 X(J)  =X(J)+X(J-K+1)
```

with N2 = $\log_2 N$. A similar FORTRAN procedure for the general solution of the linear recurrence might be coded as:

```
      DO    1   J = 1,N
      X(J)  = A(J)
    1 Y(J)  = B(J)

      DO    2      I = 1,N2
      K     = 2**(I-1)+1
      DO    2      J = K,N
      X(J)  =   X(J)+X(J-K+1)*Y(J)
    2 Y(J)  =   Y(J)*Y(J-K+1)
```

A complete derivation of these general recursive doubling procedures may be found in |197|. Stone also demonstrated an efficient implementation of the algorithm for the non-linear recurrence (5.7) for the LU-decomposition.

Since

$$d_1 = 1/a_1$$

(5.16) $$d_i = 1/(a_i - b_i\, c_{i-1}\, d_{i-1}) \ , \ i = 2,3,\ldots,n$$

is a continued-fraction expansion, we cast it into a linear form suitable for a recursive doubling algorithm. Every continued-fraction expansion is associated with a linear second-order recurrrence relation |197|. We let

$$d_i = q_{i-1}/q_i$$

in (5.16) and rearrange, obtaining the linear second-order, recurrence relation:

$$q_i = a_i\, q_{i-1} - b_i\, c_{i-1}\, q_{i-2} \quad , \; i = 2,3,\ldots,n$$

with $q_0 = 1$ and $q_1 = a_1$. This equation can be expressed as

$$Q_i = G_i * Q_{i-1} \quad , \quad i = 2,3,\ldots,n$$

where:

$$Q_1 = \begin{pmatrix} a_1 \\ 1 \end{pmatrix} \; , \; Q_i = \begin{pmatrix} q_i \\ q_{i-1} \end{pmatrix} \quad \text{and} \quad G_i = \begin{pmatrix} a_i & -b_i c_{i-1} \\ 1 & 0 \end{pmatrix} \; .$$

This is a linear, first-order recurrence similar to (5.10), except that the coefficients are now matrices. This recurrence can therefore also be solved using recursive doubling as explained in |197| in more detail.

6. Iterative Methods for a Model Problem

In this section we will discuss briefly the implementation of various iterative algorithms on vector computers. It is assumed that the reader already has some familiarity with the use of basic, iterative methods, such as presented for instance in the book of Hageman and Young |224|. Our discussion is closely tied to the CRAY-1 and the CYBER 205, but the philosophy of implementation is relevant for any vector computer.

Consider the following well-known model problem in a rectangular region G:

$$u_{xx} + u_{yy} = 1 \ , \quad (x,y) \in G$$

(6.1)
$$u(x,y) = \frac{1}{4}(x^2+y^2) \ , \quad (x,y) \in \dot{G}$$

where $G = \{(x,y)/0 < x < 1, \ 0 < y < 1\}$ and $\dot{G}$ is the boundary of G .

Discretization of the second derivatives u_{xx} and u_{yy} by the second order accurate, central differences on an N*N-grid with grid spacing H = 1/(N-1) leads us to the set of algebraic equations:

(6.2)
$$-4*U_{ij}+U_{i-1,j}+U_{i+1,j}+U_{i,j-1}+U_{i,j+1} = H^2$$

for i,j = 2,3,...,N-1. It is not our aim to give a detailed description of individual algorithms but to emphasize some important points of their implementation. In this connection, we consider three classes of algorithms, namely:

- algorithms without interdependencies of neighbouring gridpoints, such as point Jacobi and point Jacobi with conjugate gradient acceleration

- algorithms with pointwise interdependencies such as Gauss-Seidel or successive overrelaxation (SOR)

- group iterative methods such as line SOR and the alternating direction implicit (ADI) method.

In some detail, we shall demonstrate the implementation of the following subroutines on the IBM-3081K and the CRAY-1S of the DFVLR in Oberpfaffenhofen near Munich, and on the CYBER 205 at the University of Bochum:

JAC	-	point Jacobi
JACCG	-	JAC with conjugate gradient acceleration
SOR	-	point successive overrelaxation
SORRB	-	red-black or checker-board SOR
LSOR	-	line SOR
LSORZ	-	ZEBRA LSOR
VLSORZ	-	LSORZ with simultaneous treatment of the Thomas algorithms
VLSORI	-	LSORZ with iterative treatment of the Thomas algorithms
VLSORC	-	LSORZ with cyclic reduction for the Thomas algorithms
ADIR	-	alternating direction implicit with line Jacobi and relaxation.

Here, we restrict ourselves to rectangular regions, which occur rather often in practice when a physical domain D is mapped onto a mathematical region where the computations then take place. The circular area for example may be mapped onto a rectangle (see Fig. 6.1) by the well-known transformation $x = r*\cos\varphi$, $y = r*\sin\varphi$.

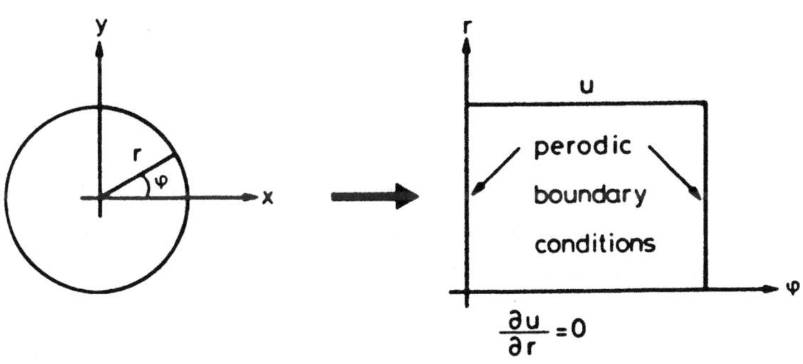

Fig. 6.1. Transformation of a circular area onto a rectangle.

More complicate examples often arise in computational physics. The flow field in Fig. 1.3 where, for instance, the computational region in the physical plane is bounded by the bow-shock and the bodysurface, may also be mapped onto a rectangle |245|. In three dimensions the mathematical domain, where the calculations will take place, is a three-dimensional cube.

6.1 Jacobi Type Methods

Methods of this type are ideal for running on vector computers. There is no need for re-structuring. Similar algorithms arise in the numerical solution of parabolic differential equations (e.g. |9|) by explicit difference methods. But as with Jacobi's method, convergence is generally so slow that these methods are not competitive. Jacobi semi-iteration based on Chebyshev polynomials |224| resp. superstep methods for the solution of parabolic differential equations |222| may be a useful alternative. Sometimes, preconditioning may be valuable (cf. |13|,|96|,|128|,|134|). In this context, we discuss only Jacobi's iterative method as a fundamental representative for more general explicit algorithms.

To begin with, the following iterative procedure can be used to solve equation (6.2):

(6.3) $$U_{ij}^{(\nu+1)} = 0.25 * (U_{i-1,j}^{(\nu)} + U_{i+1,j}^{(\nu)} + U_{i,j-1}^{(\nu)} + U_{i,j+1}^{(\nu)} - H^2) \text{ for } i,j = 2,3,\dots,N-1.$$

When i,j = 1 or i,j = N, the U_{ij} are known from the boundary conditions. (ν) is the iteration level. Here no interdependency arises between the unknowns of the new iteration level $\nu + 1$.

Comparing equation (6.3) for the grid point (I,J) with the one at (I+1,J+1), we find, that the same addition U(I+1,J) + U(I,J+1) takes place in both equations, so that it can be calculated in a previous step. The same is true for (I,J) and (I-1,J-1) and the addition U(I,J-1) + U(I-1,J). In FORTRAN this would be coded as:

```
      DO    1   J = 2,N
      DO    1   I = 2,N
    1 V(I,J)  = U(I-1,J) + U(I,J-1)

      DO    2   J = 2,N1
      DO    2   I = 2,N1
    2 U(I,J)  = 0.25 * (V(I,J) + V(I+1,J+1) - HSQ)
```

with N1 = N-1 and HSQ = H*H.

To get long and contiguously stored vectors for the CYBER 205, we use BIT-vectors representing the pattern of the loops no. 1 and 2 (see section 3.3):

```
    BIT1D = Q8VMKO (N-1,N; BIT1D)
    BIT2D = Q8VMKO (N-2,N; BIT2D)
```

with N-1 times the "1" bit and one "0" bit and so on, and similarly for BIT2D. Then, a CDC program may be of the form:

```
    WHERE (BIT1D)
    V(2,2; NN1) = U(1,2; NN1) + U(2,1; NN1)
    END WHERE

    WHERE (BIT2D)
    U(2,2; NN2) = 0.25 * (V(2,2; NN2) + V(3,3; NN2) - HSQ)
    END WHERE
```

with NN1 = N*(N-1)-1 and NN2 = N*(N-2)-2. In the case we have to implement special boundary conditions such as Neumann or symmetry conditions e.g. on the lower part of the boundary (y=0), updating of the unknown variables U(I,J) for I=1 and all J is necessary at each iteration level. However, we may use the gather/scatter instructions recommended in section 3.3. Since the first line of matrix U is needed as a vector, the elements to be gathered are separated N places from each other in physical store. Thus the periodic versions of the GATHER/SCATTER instructions can be used with a stride of N, the dimension of the matrix. In CYBER 205 FORTRAN this is coded as:

```
U1(1;N)      = Q8VGATHP (U(1,1;NN1), N,N; U1(1;N))
U1(1;N)      = ... arithmetic ...
U(1,1;NN1) = Q8VSCATP (U1(1;N), N,N; U(1,1;NN1))
```

On some vector processors such as Hitachi's S9/IAP, the following indexing will be vectorized automatically:

```
  IND = -N+1
  DO   1  K = 1,NSQ
  IND = IND + N
1 U2(IND) = ...arithmetic (K) ...
```

where in a previous step, the two-dimensional array U has been converted into the one-dimensional array U2 of length NSQ=N*N.

Indeed, no severe implementation problems arise with Jacobi's method in up-dating the variables pointwise, one after the other. On the other hand, though, convergence is generally very slow. We therefore have to look for procedures that accelerate the point Jacobi algorithm. One well-known method is the conjugate gradient acceleration. The conjugate gradient method was devised by Hestenes et al.(see in|221|) and is used since recent years as a means of accelerating otherwise slowly convergent, iterative procedures. In the absence of rounding errors, it converges to the true solution of the linear system in a finite number of iterations, the method therefore being classified as the direct type. However, when solving large systems and rounding errors are present, the direct method degenerates to an iterative one, but whose efficiency is similar to the optimum, successive over-relaxation method and the Chebyshev accelerated Jacobi method |224|. Moreover, no parameter estimates are required, in contrast with these other methods. Vectorization of the conjugate gradient method has been discussed e.g. in |128|,|134| and |183|.

The procedure we use here is a three-term form of the conjugate gradient method due to Engeli et al. |221|. It is based upon the method of steepest descend, where the quadratic form:

$$F(u) = 0.5 * (u,Au) - (f,u)$$

is minimized instead of the equivalent problem of directly solving Au = f.

In matrix notation, the Jacobi method may be expressed as:

$$U^{(\nu+1)} = G\ U^{(\nu)} + k$$

where $k = D^{-1}f$ and G is the Jacobi-iteration matrix defined by:

$$G = D^{-1}(L+U) = I - D^{-1}A.$$

The symmetric and positive definite matrix A is split into the sum of a diagonal, lower and upper triangular matrix:

$$A = D - L - U.$$

The conjugate gradient acceleration applied to Jacobi's method now is

initial step: given $u^{(0)}$

$$\rho_0 = 1$$

$$\delta^{(0)} = D^{-1}(f - Au^{(0)})$$

$$T_0 = (D^{1/2}\delta^{(0)}, D^{1/2}\delta^{(0)})$$

$$T_2 = (D^{1/2}\delta^{(0)}, D^{1/2}(I - G)\delta^{(C)})$$

$$\gamma_0 = T_0 / T_2$$

$$U^{(1)} = \gamma_0\,\delta^{(0)}$$

$$
\begin{aligned}
\text{iteration:}\quad \delta^{(v)} &= D^{-1}(f-Au^{(v)})\\
T_1 &= (D^{1/2}\delta^{(v)},\, D^{1/2}\delta^{(v)})\\
T_2 &= (D^{1/2}\delta^{(v)},\, D^{1/2}(I-G)\delta^{(v)})\\
\gamma_1 &= T_1/T_2\\
\rho_1 &= (1-\gamma_1 T_1/(\gamma_0 T_0 \rho_0))^{-1}\\
u^{(v+1)} &= \rho_1(\gamma_1\delta^{(v)}+u^{(v)})+(1-\rho_1)u^{(v-1)}\\
\gamma_0 &= \gamma_1,\quad T_0 = T_1,\quad \rho_0 = \rho_1
\end{aligned}
$$

where ρ_1 and γ_1 can be evaluated using the known vectors $U^{(\nu)}$ and $U^{(\nu-1)}$; $\delta^{(\nu)}$ being the residual vector; and where T_0, T_1, T_2 are inner products, for which most vector machines offer special subroutines supported by the hardware or written in assembly language, such as the Q8SDOT (•) on the CYBER 205 and the SDOT (•) on the CRAY-1.

Obviously no interdependencies arise between neighbouring grid points, and a vectorization therefore is straightforward.

The method of cyclic, (red-black), conjugate gradient acceleration is about twice as fast as the method presented above. About three times faster is the ZEBRA conjugate gradient acceleration. We omit a discussion of these two methods, which are similar in structure (see e.g. |224), but much more complicated to implement on vector computers.

A vectorized version of the incomplete Cholesky conjugate gradient (ICCG) method developed by Meierinck and van der Vorst can be found in |75|, |111|, |207| and |208|.

The following table shows times (in seconds) for the point Jacobi method with conjugate gradient acceleration on different computers. The vector lengths on IBM and CRAY are N = 31, 63 and 127, while on CYBER they are (N-2)*N-2 = 1021, 4093 and 16 381. The convergence criterion was

ABS (UNEW(I,J) - UOLD(I,J)) $\leq$ 0.5 * 10^{-4}

and has been reached after 56, 112 and 217 iterations respectively.

Table 6.1: Times and MFLOP rates (in parantheses) for Jacobi's conjungate gradient acceleration.

	vector length 31		63		127	
IBM-3081 K	0.430	(2.63)	3.39	(2.74)	27.6	(2.66)
CRAY-1S	0.027	(41.86)	0.18	(51.86)	1.3	(56.54)
CYBER 205	0.013	(86.93)	0.10	(93.35)	0.68	(108.09)

6.2 Gauss-Seidel Type Methods

The use of only old values of the iteration level ν for the computation of the unknowns of level $\nu+1$ is not very efficient since for each inner grid point of the new iteration level, at least two neighbouring grid points have already been up-dated (including boundary points). Therefore, equation (6.3) could better be brought into the following iterative procedure:

$$(6.4)\qquad U_{ij}^{(\nu+1)} = 0.25 * (U_{i-1,j}^{(\nu+1)} + U_{i,j-1}^{(\nu+1)} + U_{i+1,j}^{(\nu)} + U_{i,j+1}^{(\nu)} - HSQ)$$

for i,j = 2,3,...,N-1. However, vector dependencies arise now for both indices i and j, e.g. $U_{ij}^{(\nu+1)} = \varphi(U_{i-1,j}^{(\nu+1)})$ for fixed j. The algorithm therefore is recursive, and on a vector computer the dependency of U_{ij} and $U_{i-1,j}$, which is still in the pipe, degenerates to a Jacobi type algorithm. In this case, though, the iterative procedure may be replaced by the following (with N odd, N1 = N-1 and N2 = N-2):

```
      DO   2   J = 2,N1,2
      DO   2   I = 2,N1,2
    2 U(I,J)   = ... arithmetic ...

      DO   3   J = 3,N2,2
      DO   3   I = 3,N2,2
    3 U(I,J)   = ... arithmetic ...

      DO   4   J = 2,N1,2
      DO   4   I = 3,N2,2
    4 U(I,J)   = ... arithmetic ...

      DO   5   J = 3,N2,2
      DO   5   I = 2,N1,2
    5 U(I,J)   = ... arithmetic ...
```

The algorithm is called red-black or checker-board pattern, and is easily implemented on vector computers. One only has to copy the original double DO-loop four times and change the range of the indices appropriately. The arrangement of the grid points for the four loops is shown in Fig. (6.2).

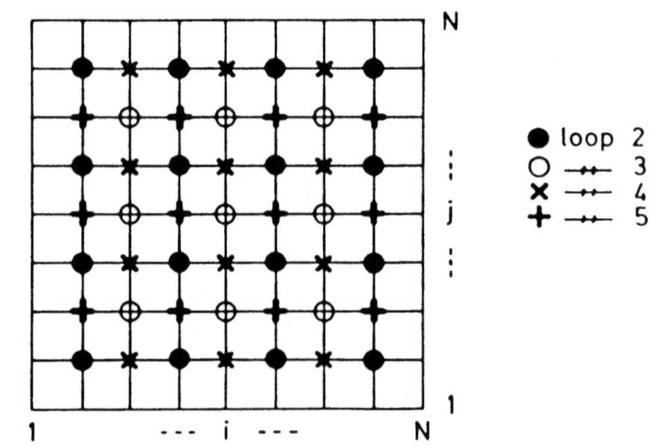

Fig. 6.2: Grid points for red-black ordering.

Since the vectors have a relatively short length (16, 32 and 64 for N = 33, 65 and 129 respectively), this algorithm is not yet very efficient on vector machines of the CYBER 205 type. To achieve long vectors of length 512, 2048 and 8192, we have to construct suitable BIT-vectors. Let BITED and BITOD be the BIT-vectors pointing to the even and odd inner points of the rectangle. Here, an inner point is called even if I+J is even. Then the essence of the algorithm could be coded as follows:

```
WHERE (BITED)
U(2,2;NN2) = 0.25 * (U(1,2;NN2) + U(3,2;NN2) + U(2,1;NN2) +
                     U(2,3;NN2) - HSQ)
END WHERE

WHERE (BITOD)
U(2,2;NN2) = 0.25 * (U(1,2;NN2) + U(3,2;NN2) + U(2,1;NN2) +
                     U(2,3;NN2) - HSQ)
END WHERE
```

with NN2 = (N-2) * N-2. This computation is not yet optimal. The NN2 components must be evaluated twice, although only about half of them are needed. Following the proposition of section 3.3, we compress all the LRED even components into the vector RED and all the LBLA odd components into the vector BLA. Then the compressed red-black variant of the kernel above for N = 65 has the form

```
WHERE (BITED)
RED(34,1;LRED) = 0.25 * (BLA(1,1;LRED) + BLA(33,1;LRED)
                 + BLA(34,1;LRED) + BLA(1,2;LRED) - HSQ)
END WHERE

WHERE (BITOD)
BLA(34,1;LBLA) = 0.25 * (RED(2,1;LBLA) + RED(34,1;LBLA)
                 + RED(35,1;LBLA)+ RED(2,2;LBLA) - HSQ)
END WHERE
```

If the matrix of the system has some special properties (see e.g. Young |224|),the Gauss-Seidel method can be accelerated by a real number ω with $1 \leq \omega < 2$, known as the relaxation factor, and the method is called successive overrelaxation (SOR), resulting in the weighted relation:

$$(6.5)\quad U_{ij}^{(\nu+1)} = (1-\omega)U_{ij}^{(\nu)} + 0.25\omega * (U_{i-1,j}^{(\nu+1)} + U_{i,j-1}^{(\nu+1)} + U_{i+1,j}^{(\nu)} + U_{i,j+1}^{(\nu)} - HSQ)$$

for i,j = 2,3,...,N-1. For our model problem, the optimum value of ω, denoted by ω_b, is given by:

$$\omega_b = 2/(1 + \sin \pi h).$$

This method with optimum ω_b has been implemented on the computers listed below. For SOR, the red-black SORRB and the compressed red-black SORRBC (for CYBER only), we give times and MFLOP rates in Table 6.2.

Table 6.2: Times and MFLOP rates (in parantheses) for SOR and red-black SOR.

	SOR	SORRB	SORRBC
N = 31 :			
IBM-3081 K	0.174 (2.96)	0.126 (2.87)	
CRAY-1S	0.059 (8.73)	0.011 (32.85)	
CYBER 205	0.047 (10.96)	0.0064 (56.46)	0.0050 (72.27)
N = 63:			
IBM-3081 K	1.43 (2.91)	1.02 (2.79)	
CRAY-1S	0.45 (9.24)	0.065 (43.8)	
CYBER 205	0.42 (9.9)	0.044 (64.7)	0.031 (91.83)
N = 127:			
IBM-3081 K	11.97 (2.78)	8.31 (2.69)	
CRAY-1S	3.61 (9.23)	0.42 (53.15)	
CYBER 205	3.97 (8.39)	0.33 (67.64)	0.221 (101.0)

The MFLOP rates have been decreased somewhat owing to the testing for convergence for the simple model problem assuming a large portion of the overall computation. The simpler the arithmetic of the iterative algorithm, the more expensive is pointwise testing for convergence on vector computers. We therefore computed UOLD - UNEW only in the centre point of the region G. After the required accuracy has been reached at this grid point, the convergence test has been carried out in the whole region.

As can be seen from Table 6.2, on serial computers SORRB is also faster than the natural ordered SOR. Although the asymptotic rates of convergence are the same for both methods (cf. Alefeld |1|), the average rate of convergence can be significantly less than the asymptotic rate (see e.g. Miles |232|).

6.3 Group Iterative Methods

For group or block iterative methods, groups or blocks of unknowns are improved simultaneously. The blocks of unknowns to be improved simultaneously are determined by an partitioning imposed on the coefficient matrix. Examples are line Jacobi (LJAC), successive line over-relaxation (LSOR) and alternating direction implicit (ADI) methods. We discuss the implementation of this class of algorithms while using the LSOR method described e.g. in |224|. We start with systems

(6.6) $$A_{jj} U_j^{(\nu+1)} = P_j \quad ,$$

which result from the simultaneous solution of the equations along grid lines y_j = const. of the discrete region. Here

$$P_j = (1-\omega)A_{jj} U_j^{(\nu)} + \omega\{ U_{j-1}^{(\nu+1)} + U_{j+1}^{(\nu)} - HH\} ,$$

where

$$A_{jj} = \begin{pmatrix} 4 & -1 & & \\ -1 & 4 & -1 & \\ \ddots & \ddots & \ddots & \\ & -1 & 4 \end{pmatrix} ,$$

HH means $h^2 I$ with the identity matrix I and U_j and P_j are vectors at iteration level ν resp. $\nu+1$. The optimum value of the relaxation factor for the model problem is:

$$\omega_D = \frac{2(1+2a^2)}{(1+\sqrt{2}a)^2}$$

with $a = \sin \pi h/2$. We solve the systems (6.6) by variants of the Thomas-algorithm discussed in section 5.4. The first problem is that $U_j^{(\nu+1)}$ is a function of $U_{j-1}^{(\nu+1)}$ which may be still in the pipe when it is needed; so, to avoid interdependencies of neighbouring grid lines, a ZEBRA pattern is introduced resulting in an even-odd structure of the lines y_j = const. within each iteration step:

```
  J1 = 2  ,  J2 = N1
  DO   1     K = 1,2
  DO   2     J = J1,J2,2
  DO   2     I = 2,N1
  P(I,J)  = ...
2 CONTINUE
  J1 = 3  ,  J2 = N2
1 CONTINUE
```

with N1 = N-1, N2 = N-2. For K=1 (K=2), the Thomas algorithms are carried out for all lines with even (odd) indices. Unfortunately the Thomas-algorithms contain three highly serial recursions, as pointed out in the preceding chapter. But if we simultaneously solve all the "black" systems, and afterwards all the "white" systems, the effectiveness of the algorithm is saved. All that is left to do is to interchange I and J. For the example, this results in:

```
  I1 = 2  ,  I2 = N1
  DO   1     K = 1,2
  DO   2     J = 2,N1
  DO   2     I = I1,I2,2
2 P(I,J) = ...
  I1 = 3  ,  I2 = N2
1 CONTINUE
```

Without the ZEBRA pattern and for ω= 1, this algorithm degrades to Line Jacobi . The interchange of I and J is necessary to run over the first index of P as it is stored in memory.

Another method to by-pass the recurrences in ZEBRA-SLOR is to iterate these recurrences by the odd-even acceleration |204| of Traub. For example, we look at the LU-decomposition (5.7):

```
      DO     1   J = 3,N1
      D(J)   = 1./(A(J)-E(J)*D(J-1))
    1 CONTINUE
```

A trivial manipulation of this loop gives the iterative evaluation of the vector D:

```
      DO     2   J = 2,N1
    2 D(J)   = 1./A(J)
      DO     1   K = 3,JEND
      DO     1   J = K,N1
      D(J)   = ...
    1 CONTINUE
```

where the number of iterations JEND depends on the dominance of the diagonals |81|.

It should be noted that the even-subscripted D(J) only depend on the odd-subscripted ones and vice versa. The next step therefore is the implementation of an odd-even structure for Traub's iterative treatment of the recurrences:

```
      DO    1   K = 3,JEND
      DO    3   J = 3,N2,2
    3 D(J)  = ...
      DO    4   J = 4,N1,2
    4 D(J)  = ...
    1 CONTINUE
```

Another method for solving linear recurrences is the method of cyclic reduction explained in detail in section 7.3. The implementation of cyclic reduction on vector computers is staightforward. Following formulae (5.12) to (5.15), the forward substitution

```
      DO   1   I  = IF,IL
    1 G(I) = F(I) - GAM(I)*G(I-1)
```

leads us to the following general cyclic reduction algorithm for first-order linear recurrences for N = 2**M:

```
      DO    4    I = IF,IL
    4 GA1(I) =  GAM(I)
      DO    3    L = 1,M
      K  = 2**(M-1) + (IF-1)
      DO    2    I = K,IL
      I1 = I-K+(IF-1)
      G1(I) = F(I)-GA1(I)*G(I1)
    2 A1(I) = GA1(I)*GA1(I1)
      DO    1    I = K,IL
      G(I)   = G1(I)
    1 GA1(I) =   A1(I)
    3 CONTINUE
```

There are no further interdependencies between the left and right hand sides. Loop 2 contains only vector operations. When A is strict diagonally dominant, M might even be smaller depending on the required accuracy.

The ADI type methods also belong to the class of block iterative methods. The first half step of the iteration simultaneously solves a tridiagonal system for each row of the grid, and the second half step does the same for each column. If in the first half step the storing takes place with a natural ordering, the second does not. In this case and for long vectors, excessive paging on machines with virtual memory may occur, so that for the second half step the implementation of a previously discussed algorithm will be necessary. This is demonstrated next using one of the most famous ADI type methods, the Peaceman-Rachford method which is based on the matrix splitting:

$$A = H + V ,$$

where A, H and V are assumed to be symmetric and positive definite. The problems of vectorization are the same as in all methods based on operator splitting (cf. |44|,|45| and |161|). The method (cf. |224|) is defined by:

$$(H+\rho_n I)u^{(n+1/2)} = f - (V-\rho_n I)u^{(n)}$$

$$(V+\rho_n' I)u^{(n+1)} = f - (H-\rho_n' I)u^{(n+1/2)}$$

with positive relaxation parameters ρ_n, ρ_n'. Typically H and V are tridiagonal matrices and the system again can be solved using the Thomas algorithm.

For the model problem (6.1), the right hand side of the first half step of the iteration is calculated by means of the DO-loop:

```
      DO     1    J = 2,N1
      DO     1    I = 2,N1
    1 P(I,J)  = - HSQ + U(I,J-1) + U(I,J+1) - RHO*U(I,J)
```

with N1 = N-1, HSQ = H * H and the relaxation parameter RHO, whereas the right hand side of the second half step of the iteration is calculated by the FORTRAN code:

```
      DO     2    I = 2,N1
      DO     2    J = 2,N1
    2 P(I,J)  = - HSQ + U(I-1,J) + U(I+1,J) - RHO*U(I,J)
```

The simultaneous evaluation of the Thomas algorithms will be auto-vectorized by the CRAY compiler. On the CYBER, the first half step with natural ordering also vectorizes easily. However, in the second half iteration step, the vectors to be up-dated are not stored contiguously in memory; they are stored row-wise. In this case, however, the solution is to gather all the rows I = const. before performing the arithmetic.

In the following Table 6.3 we compare several group iterative methods discussed in this section, and compare them with the line Jacobi (LJAC) method for vector lengths N = 31, 63 and 127. The methods considered here are the non-vectorized successive line over-relaxation (LSOR), LSOR with ZEBRA pattern (LSORZ), the fully vectorized version of LSOR with ZEBRA pattern, and simultaneous solution of the Thomas algorithms

(VLSORZ), LSORZ with iterative treatment of the Thomas algorithms due to Traub (VLSORI), LSORZ with cyclic reduction for the Thomas algorithms (VLSORC) and the alternating direction implicit method with relaxation (ADIR).

Table 6.3: Times (in seconds) for group iterative methods.

	IBM-3081 K			CRAY-1S			CYBER 205		
	31	63	127	31	63	127	31	63	127
LJAC	1.47	17.66	155.7	0.15	1.36	11.76	0.205	1.61	11.0
LSOR	0.24	2.00	14.9	0.058	0.44	3.34	0.066	0.46	3.35
LSOR Z	0.18	1.43	10.9	0.046	0.33	2.53	0.050	0.35	2.53
VLSOR Z	0.18	1.45	11.6	0.015	0.08	0.54	0.031	0.07	0.36
VLSOR I	0.51	4.10	31.5	0.067	0.36	2.18	0.074	0.36	1.85
VLSOR C	0.90	7.55	51.1	0.100	0.59	3.75	0.128	0.68	3.76
ADIR	0.37	2.94	25.2	0.030	0.16	1.09	0.032	0.16	1.02

The vector lengths in LJAC, LSOR, LSORZ, VLSORI, VLSORC and ADIR are 31, 63 and 127 and for VLSORZ 16, 32 and 64 respectively. It is important to note, that the CYBER 205 reaches half of its maximum speed for vectors of length of about 100, and 90 per cent of its maximum speed for vectors with length of about 1000, which is far from being attained in the above mentioned programs. Vector lengths of 16 and 32 result in performance rates of about 15 resp. 25 per cent of the maximum performance for a two-pipe CYBER 205, while for the CRAY-1S these numbers are 70 resp. 85 per cent of the maximum.

7. Vectorization of Simple Numerical Algorithms

Before applying the ideas developed in chapters 3 to 6 to complete production codes, we first discuss the basic structure of simple numerical algorithms and their implementation on vector computers. The problems selected are

Polynomial evaluation
Computation of the largest eigenvalue of a matrix
Cyclic reduction
Systems of non-linear equations
Runge-Kutta methods for systems of ordinary differential equations
Explicit superstep methods for parabolic equations
Finite difference smoothing.

Most of the problems presented have a Jacobi type structure (see section 6.1) and therefore are easily vectorized. One important point, however, will not be discussed here: the change in stability and convergence properties when restructuring the algorithms. For the red-black SOR of section 6.2, for example, Alefeld |1| derived the same asymptotic convergence rate as the SOR with natural ordering. Roensch |163| applied the rounding error analysis of Stummel |243| and Larson and Sameh |232| to a variety of basic arithmetic expressions and more complex algorithms such as Horner's scheme, especially for the CRAY-1.

7.1 Polynomial Evaluation

Horner's method is regarded as a useful algorithm for evaluating algebraic polynomials. Unfortunately, for only one evaluation, the method uses a recurrence relation. However, if we wish to compute the values of a polynomial:

$$P_n(x) = a_{n+1} + a_n x + a_{n-1} x^2 + \dots + a_1 x^n$$

for many values of x, this process can be vectorized efficiently (cf. |16|, |105|, |107| and |108|). To explain this in more detail, we can rewrite the polynomial in the form:

$$P_n(x) = a_{n+1}+x(a_n+x(a_{n-1}+\ldots x(a_2+xa_1)\ldots))$$

and construct the sequence:

$$\begin{aligned}
b_1 &= a_1x \\
b_2 &= (a_2+b_1)x \\
b_3 &= (a_3+b_2)x \\
&\ \ \ldots \\
b_n &= (a_n+b_{n-1})x \\
b_{n+1} &= a_{n+1}+b_n \quad .
\end{aligned} \tag{7.1}$$

It is easily verified that the final result of this recurrence relation is $b_{n+1} = P_n(x)$. If one is only interested in b_{n+1}, each b_K $(K = 1,2,\ldots,n+1)$ can be stored and then re-defined in one place: $b = b_K$. For a set of M values x_i, the b's may be calculated simultaneously. For all $i = 1,2,\ldots,M$ the above algorithm can be written in the form:

$$\begin{aligned}
b(i) &= a_1 * x(i) \\
b(i) &= (a_2+b(i)) * x(i) \\
b(i) &= (a_3+b(i)) * x(i) \\
&\ldots \\
b(i) &= a_{n+1}+b(i).
\end{aligned}$$

This procedure might be coded in FORTRAN as:

```
    DO    1   I = 1,M
    B(I) = A(1) * X(I)
    DO    2   J = 2,N
  2 B(I) = (A(J) + B(I)) * X(I)
  1 B(I) = A(N+1) + B(I)
```

where M is the number of different evaluations and N the degree of the polynomial. A simple re-structuring then results in the equivalent procedure:

```
      DO    1    I = 1,M
    1 B(I) = A(1) * X(I)
      DO    2    J = 2,N
      DO    2    I = 1,M
    2 B(I) = (A(J) + B(I)) * X(I)
      DO    3    I = 1,M
    3 B(I) = A(N+1) + B(I)
```

Since A(J) is a constant for the innermost DO-loop 2, this loop is a linked triad of the form:

(scalar + vector) * vector ,

so that this sequence gains efficiency from chaining on the CRAY. For the CYBER, with implicit descriptor notation (see section 3.3), a vectorizable procedure might be:

```
      BD = A(1) * XD
      DO   2    J = 2,N
      BD = A(J) + BD
      BD = BD * XD
    2 CONTINUE
      BD = A(N+1) + BD
```

As mentioned in section 3.3, the link instruction in loop 2 can be calculated during one cycle per result per pipe on the CYBER 205.

If one needs values of the polynomial only for selected x_i of a given array $x_1, x_2, \dots, x_M$, say:

$P(x_1)$, $P(x_{K+1})$, $P(x_{2K+1})$, ...

then, at least on the CYBER, it would be better to collect the active elements $x_1, x_{K+1}, x_{2K+1}, \dots$. For a detailed discussion of this problem, the reader is referred to Kascic |105|,|108|.

7.2 The Power Method for the Computation of the Largest Eigenvalue of a Matrix

As an application of the vectorized matrix*vector operation as explained in section 5.1, we discuss the power method, associated with the name of v. Mises, which consists in determining the dominant, characteristic root (or eigenvalue) and the corresponding eigenvector of a matrix by a simple, iterative process.

For simplicity, we shall not investigate the problem in general, but restrict ourselves to considering only the case where the eigenvalue λ_1 of maximal modulus is real and simple, i.e.

$$|\lambda_1| > |\lambda_2| \geq \ldots \geq |\lambda_n| .$$

We take an arbitrary initial vector $x^{(0)}$ and form the iteration process:

$$x^{(k)} = A\, x^{(k-1)} = A^2 x^{(k-2)} = \ldots = A^k\, x^{(0)}.$$

Expanding the vector $x^{(0)}$ in terms of the eigenvectors, one can show that the sequence $x^{(k)}$, $k = 0,1,\ldots,$ will converge to an eigenvector corresponding to λ_1. Moreover, λ_1 is approximately equal to the ratio of any corresponding components of two adjacent iterations:

$$\lambda_1 \approx \frac{x_k^{(i+1)}}{x_k^{(i)}} .$$

To carry out the iterations in practice, these ratios must be computed for several components.

Another division of the vectors by its largest component is necessary to avoid the growth of components. However, it is often sufficient to divide the vectors by the first components, so considerably simplifying the vectorization of the algorithm. The iterative process then takes the following form:

$$x^{(k)} = Ay^{(k-1)}$$

$$y^{(k)} = \frac{1}{||x^{(k)}||} * x^{(k)}$$

with $||x^{(k)}|| = |x_1^{(k)}|$, which only will cause trouble when the eigenvector has zero as its first component. A vectorized FORTRAN version including a linked triad of the form:

vector + scalar * vector ,

this being now similar to that discussed in section 5.1 for the matrix * vector operation, will be omitted.

7.3 Cyclic Reduction

The principle of cyclic reduction may also be applied to the direct solution of the general tridiagonal system of equations (5.1). This direct method, which is very suitable for vector computers (cf. |80|, |125|,|160|), was originally developed by Hockney |226| and has been used in direct solutions to Poisson's equation by Buzbee |219| and others.

The basic idea, which for linear recurrences is explained in section 5.4, is to eliminate the odd-subscripted unknowns from the even-numbered equations and then re-group the even-numbered equations together by row and column permutations to generate another tridiagonal system of lower order. We will assume for simplicity that the order n of the system is a power of 2, although the algorithm works for arbitrary n. By way of an example, we shall consider a system Au = f for n = 8 and the 8x8-matrix:

$$A = \begin{pmatrix} 2 & -1 & & & \\ -1 & 2 & -1 & & \\ & \ddots & \ddots & \ddots & \\ & & -1 & 2 & -1 \\ & & & -1 & 2 \end{pmatrix}.$$

The first step of the algorithm is to add a multiple (here 0.5) of an odd row to the even row below it

(odd)	-1	2	-1		
(even)		-1	2	-1	

so achieving the following:

(odd)	-1	2	-1		
(even)	-0.5	0	1.5	-1	

In the second step we add a multiple (here 0.5) of an odd row to the even row above it:

(even)	-0.5	0	1.5	-1	
(odd)			-1	2	-1

so obtaining the following:

(even)	-0.5	0	1	0	-0.5
(odd)			-1	2	-1

The even-numbered rows now form another tridiagonal system of equations involving only the even-subscripted variables:

$$(7.2)\qquad \begin{pmatrix} 1 & -0.5 & & \\ -0.5 & 1 & -0.5 & \\ & -0.5 & 1 & -0.5 \\ & & -0.5 & 1.5 \end{pmatrix} \begin{pmatrix} U_2 \\ U_4 \\ U_6 \\ U_8 \end{pmatrix} = \begin{pmatrix} f_2 + 0.5 \cdot (f_1 + f_3) \\ f_4 + 0.5 \cdot (f_3 + f_5) \\ f_6 + 0.5 \cdot (f_5 + f_7) \\ f_8 + 0.5 \cdot f_7 \end{pmatrix}$$

The original size of the system was $n = 2^3$. The smaller system now has size 2^2. The odd-subscripted variables can be computed directly by substituting the known values of U_2, U_4, U_6 and U_8 into the odd-numbered equations, e.g.

$$U_3 = 0.5 * (f_3 + U_2 + U_4).$$

However, the explicit solution of the 2^2-system is not necessary since the above procedure can be repeated on this smaller system:

(odd)	-0.5	1	-0.5		
(even)		-0.5	1	-0.5	

resulting in:

(odd)	-0.5	1	-0.5		
(even)	-0.25	0	0.75	-0.5	

and similarly:

(even)	-0.25	0	0.75	-0.5	
(odd)			-0.5	1	-0.5

which resulting in:

(even)	-0.25	0	0.5	0	-0.25
(odd)			-0.5	1	-0.5.

Here the smaller system of half the size has a somewhat different form owing to its smaller size:

$$(7.3) \quad \begin{pmatrix} 0.5 & -0.25 \\ -0.25 & 1.25 \end{pmatrix} \begin{pmatrix} U_4 \\ U_8 \end{pmatrix} = \begin{pmatrix} r_4 \\ r_8 \end{pmatrix}$$

with

$$r_4 = f_4+0.75*(f_3+f_5)+0.5*(f_2+f_6)+0.25*(f_1+f_7)$$

$$r_8 = f_8+0.75*f_7+0.5*f_6+0.25*f_5$$

This process continues until one gets a single equation in one unknown. Again we have to add a multiple (here 0.5) of the first row to the second row

(odd)	0.5	-0.25
(even)	-0.25	1.25

so that, after three steps, one equation in one unknown, namely U_8, remains:

$$1.125 * U_8 = r_8 + 0.5 * r_4 \quad .$$

Having determined U_8, substituting back into the other equations can proceed. U_8 is substituted into the first equation of system (7.3), and U_4 and U_8 are substituted into the first and third equation of system (7.2), so computing all the odd-subscripted variables from the original system and the algorithm is terminated.

A general derivation of cyclic reduction for systems Au = f with the tridiagonal matrix A of (5.1) is now straightforward. We carry out the same procedure described above to eliminate references to odd-subscripted variables in the even-numbered equations, using the following elementary row operations:

$$-\alpha_{2i}*\mathrm{Row}(2i-1) + \mathrm{Row}(2i) - \beta_{2i}*\mathrm{Row}(2i+1)$$

with $\alpha_{2i} = b_{2i}/a_{2i-1}$ and $\beta_{2i} = c_{2i}/a_{2i+1}$. Then the modified tridiagonal system of equations becomes:

$$-(b_{2i-1}\alpha_{2i}) * U_{2i-2} + (a_{2i} - c_{2i-1}\alpha_{2i} - b_{2i+1}\beta_{2i})U_{2i} - c_{2i+1}\beta_{2i} * U_{2i+2} =$$

$$= f_{2i} - \alpha_{2i}f_{2i-1} - \beta_{2i}f_{2i+1} \qquad , \text{ for } i = 1,2,\ldots,2^{K-1}$$

where $b_1 = c_n = 0$. In general, each step of cyclic reduction reduces a 2^k*2^k-system to one of size $2^{k-1}*2^{k-1}$, and after k steps we obtain one equation for the unknown U_2. Now substitution similar to that described above, yields the final result.

A vectorization of the reduction steps is now straightforward if the original matrix A is stored by diagonals. An implementation of the algorithm on vector computers is discussed in detail in |125|.

7.4 Systems of Non-linear Equations

We now proceed to a short description of iterative methods for the solution of systems of n non-linear equations (see e.g. |18|,|216|, |237| and |244|) of the form:

$$f_1(x_1,x_2,...,x_n) = 0$$
$$f_2(x_1,x_2,...,x_n) = 0$$
$$...$$
$$f_n(x_1,x_2,...,x_n) = 0$$

which is also written in vector notation as:

$$f(x) = 0$$

with $f = (f_1,f_2,...,f_n)^T$ and $x = (x_1,x_2,...,x_n)^T$. In general, the solution of non-linear systems for large n presents some severe difficulties concerning the correct initial data and convergence towards an unique solution x^* (if it exists). Let the above system be transformed into the equivalent form:

$$x = g(x) \quad .$$

One well-known, non-linear iteration method then is the generalized Newton method:

$$x^{(k+1)} = g(x^{(k)}) = x^{(k)} - J^{-1}(x^{(k)}) * f(x^{(k)})$$

where J(x) is the Jacobian matrix defined by

$$(J(x))_{ij} = \partial f_i(x)/ \partial x_j \quad .$$

It can be shown by means of a Taylor's series expansion about the exact solution x^* that under the circumstances that if f(x) has continuous second partial derivatives and the Jacobian matrix is nonsingular at x^* and if the iteration process determined by the generalized Newton method converges at all, it converges quadratically, |236|.

Instead of evaluating $J(x^{(k)})$ and computing the inverse at each iteration step, we solve the equivalent sequence of linear systems

$$J(x^{(k)})\ x^{(k+1)} = J(x^{(k)})\ x^{(k)} - f(x^{(k)})$$

provided the Jacobian is non-singular. The matrix $J(x^{(k)})$ and the right-hand side are known. These systems must be solved by direct or iterative linear system solvers. Vectorization then takes place according to the procedures described in section 5.4 and chapter 6 and therefore will not be discussed further here.

In practice the Jacobian $J(x)$ of $f(x)$, which should be evaluated at $x^{(k)}$, will not be re-evaluated at each iteration, but only when necessary to ensure convergence. That is, if the iteration is converging with fixed J, the time-consuming process of re-evaluating and triangulating J is omitted.

As well as other difficulties, the speed of convergence of Newton's method depends on a good approximation to the solution being available to start the iteration. The iterative techniques can be improved in many ways, however. One is the use of componentwise monotonically convergent iterations as proposed e.g. by Toernig et al.in |216|,|244|, where a general class of monotone convergent iterative processes of the Gauss-Seidel type has been discussed.

To illustrate the vectorization of this class of methods, let

$$x^{(k)i} = (x_1^{(k+1)}, \ldots, x_{i-1}^{(k+1)}, x_i^{(k)}, \ldots, x_n^{(k)}) .$$

An iterative scheme of the form:

$$x^{(k+1)} = x^{(k)} - P^k(x^{(k)i}) \cdot f(x^{(k)i}) \quad , \quad k = 0,1,\ldots$$

for $i = 1,2,\ldots,n$ is called to be of Gauss-Seidel type. The nxn matrix P^k depends on the algorithm used. In the important case of diagonal matrices P^k with diagonal elements P_i^k, the iterative scheme can be written componentwise as:

$$x_i^{(k+1)} = x_i^{(k)} - P_i(x^{(k)i}) f_i(x^{(k)i}) \quad , \quad k = 0,1,\ldots$$

for all indices $i = 1,2,\dots,n$. If

$$J(x) = D(x) - L(x) - U(x)$$

is any partition of the Jacobian into diagonal, strictly lower and strictly upper triangular parts, respectively, we may set:

$$P^{(k)}(x) = D^{-1}(x)$$

and hence

$$P_i^k(x) = \frac{1}{d_i(x)} = \left(\frac{\partial f_i(x)}{\partial x_i}\right)^{-1} \quad , \quad i = 1,2,\dots,n \; .$$

Let f be an off-diagonally antitone and strictly diagonally isotone mapping (see e.g. |236|) and there exist points $x^{(0)}$ and $y^{(0)}$ satisfying

$$x^{(0)} \leqslant y^{(0)} \quad , \qquad f(x^{(0)}) \leqslant 0 \leqslant f(y^{(0)})$$

componentwise. Moreover, if $P^k(x)$ are positive functions, then the iterates $x^{(k)}$ and $y^{(k)}$ converge monotonically to x^* and y^*, respectively, with

$$x^{(0)} \leqslant x^* \leqslant y^* \leqslant y^{(0)} \quad , \quad f(x^*) = f(y^*) = 0.$$

The main advantage of this iterative process is the global convergence criterion given by any norm of the vectors $x^{(k)}-y^{(k)}$.

As an example, consider the stationary Burger's equation:

$$\frac{1}{2}(U^2)_x = \mu U_{xx}$$

with a diffusion term on the right-hand side. The convective term on the left-hand side of the equation may be discretized at the mesh point x_i by

$$(V_i^2)_x \doteq \frac{1}{\Delta x}(V^2_{i+\frac{1}{2}} - V^2_{i-\frac{1}{2}}) = \frac{1}{\Delta x}\left\{\left(\frac{V_{i+1}+V_i}{2}\right)^2 - \left(\frac{V_i+V_{i-1}}{2}\right)^2\right\}$$

so that the non-linear system of equations takes the form

$$f_i(V) = \frac{\alpha}{2}(V_{i+1}^2 + 2V_i(V_{i+1} - V_{i-1}) - V_{i-1}^2) - V_{i+1} + 2V_i - V_{i-1} = 0$$

for $i = 2,3,\ldots,n-1$, $V = (V_1, V_2, \ldots, V_n)^T$ and $\alpha = \Delta x/4\mu$. It can be shown that the Jacobian of the system is a diagonally dominant L-matrix (see e.g. |224|). The positive diagonals of the Jacobian are computed to be:

$$d_i(V) = 2 + \alpha(V_{i+1} - V_{i-1})$$

The iterative scheme then becomes:

$$\begin{aligned} V_i^{(k+1)} &= g(V_{i-1}^{(k+1)}, V_i^{(k)}, V_{i+1}^{(k)}) \\ &= V_i^{(k)} - \left\{ \frac{\alpha}{2}\left((V_{i+1}^{(k)})^2 + 2V_i^{(k)}(V_{i+1}^{(k)} - V_{i-1}^{(k+1)}) - (V_{i-1}^{(k+1)})^2\right) \right. \\ &\quad \left. - V_{i+1}^{(k)} + 2V_i^{(k)} - V_{i-1}^{(k+1)} \right\} \Big/ \left(2+\alpha(V_{i+1}^{(k)} - V_{i-1}^{(k+1)})\right) \end{aligned}$$

for $i = 2,3,\ldots,n-1$. In this form the scheme is highly recursive owing to the dependency between $V_i^{(k+1)}$ and $V_{i-1}^{(k+1)}$. However, a vectorization is possible when applying the red-black re-ordering of section 6.2. In detail, we have two stages:

$$V_i^{(k+\frac{1}{2})} = g(V_{i-1}^{(k)}, V_i^{(k-\frac{1}{2})}, V_{i+1}^{(k)})$$

$$V_i^{(k+1)} = g(V_i^{(k+\frac{1}{2})}, V_{i+1}^{(k)}, V_{i+2}^{(k+\frac{1}{2})})$$

for $i = 2,4,\ldots,n-2$ and n even, which may be evaluated independently. At each stage, the values of the right-hand side are known.

Under certain assumptions which are similar to those for linear, red-black ordering, convergence may be proven (see e.g. Alefeld |1| and Hageman et al. |224|). However, the author has no knowledge whether or not the profitable monotone convergence property will be preserved.

7.5 Runge-Kutta Time-Stepping Methods for Systems of Ordinary Differential Equations

Runge-Kutta methods have a unique place amongst the classical discrete variable methods for solving the initial value problem of ordinary differential equations (cf. |228|, |230|, |239|, |240|):

$$\frac{dy}{dx} = y' = f(x,y) \ , \ x \in [a,b]$$
$$y(a) = y_0$$

where $y = (y^1(x), y^2(x), \ldots, y^s(x))^T$, and the initial value y_0 is given. The idea is to construct a formula

$$y_{n+1} = y_n + h\phi(x_n, y_n, h) \ , \ n = 0,1,2,\ldots$$

which agrees with the Taylor series expansion of y as closely as possible without involving derivatives of f. An explicit formula of second-order accuracy which requires two evaluations of f per step is (for s = 1):

$$y_{n+1} = y_n + \frac{h}{2}\left\{f(x_n, y_n) + f(x_n+h, y_n+hf(x_n,y_n))\right\}$$

which was first obtained by Heun |228|.

The form of the methods is ideal for practical computations. They do not require any special starting procedure, in contrast to multistep methods such as Adams-Bashford method and allow easy change of stepsize. The most well-known formula is perhaps the 4-stage fourth order accurate Runge-Kutta method which, for s = 1, takes the following form:

$$y_{n+1} = y_n + \frac{1}{6}(k_1+2k_2+2k_3+k_4)$$

where:

$$k_1 = hf(x_n, y_n) \ , \quad k_2 = hf(x_n + \frac{h}{2}, y_n + \frac{k_1}{2})$$

$$k_3 = hf(x_n + \frac{h}{2}, y_n + \frac{k_2}{2}) \ , \quad k_4 = hf(x_n+h, y_n+k_3) \ .$$

For $s = 1$ the method is highly recursive and therefore not vectorizable. For $s > 1$, however, each Runge-Kutta stage may be evaluated simultaneously for $y^1, y^2, \ldots, y^s$. The larger the number s of unknowns, the faster the vector algorithm.

The application of the method to partial differential equations is now straightforward and leads to a large system of ordinary differential equations. By way of an example, we discuss the implementation of the classical 4-stage Runge-Kutta scheme for the model problem:

$$U_t = U_{xx} \quad , \; x \in [0,1] \; , \; t > 0$$

$$U(0,t) = U_0(t) \quad , \; U(1,t) = U_1(t)$$

$$U(x,0) = g(x)$$

on vector computers. A method that has been studied extensively, is the "method of lines", also called semi-discrete approximation |227|. The partial differential equation is reduced to a system of ordinary differential equations by simply replacing the spatial derivative of the right-hand side by finite differences. Let:

$$PV = \frac{1}{\Delta x^2} \begin{pmatrix} 2 & -1 & & & \\ -1 & 2 & -1 & & \\ & \ddots & \ddots & \ddots & \\ & & -1 & 2 & -1 \\ & & & -1 & 2 \end{pmatrix} V$$

be the discretization of the right-hand side of the differential equation in the discrete interval |0,1| with step-size Δx and $V = (V_1, V_2, \ldots, V_s)^T$. The system of equations then results in the so-called semi-discrete representation:

$$\frac{dV}{dt} + PV = 0 \; .$$

The 4-stage Runge-Kutta scheme for the solution of this equation on time level n then becomes:

$$W^{(0)} = V^{(n)}$$

$$W^{(1)} = W^{(0)} - \frac{1}{2}\Delta t \, PW^{(0)}$$

$$W^{(2)} = W^{(0)} - \frac{1}{2}\Delta t \, PW^{(1)}$$

$$W^{(3)} = W^{(0)} - \Delta t \, PW^{(2)}$$

$$W^{(4)} = W^{(0)} - \frac{1}{6}\Delta t \, (PW^{(0)} + 2PW^{(1)} + 2PW^{(2)} + PW^{(3)})$$

$$= \frac{1}{3} (W^{(1)} + 2W^{(2)} + W^{(3)} - W^{(0)} - \frac{\Delta t}{2} PW^{(3)})$$

$$V^{(n+1)} = W^{(4)} .$$

In |230|, the scheme has been used to integrate the hyperbolic Euler equations for inviscid flow in time. There another multistage, two-level scheme of the Runge-Kutta type is revived, the Gary scheme (1964). Again let n denote the time level ($t = n \cdot \Delta t$), we have:

$$W^{(0)} = V^{(n)}$$

$$W^{(1)} = W^{(0)} - \Delta t \, PW^{(0)}$$

$$W^{(2)} = W^{(0)} - \frac{1}{2}\Delta t \, (PW^{(0)} + PW^{(1)})$$

$$W^{(3)} = W^{(0)} - \frac{1}{2}\Delta t \, (PW^{(0)} + PW^{(2)})$$

$$V^{(n+1)} = W^{(3)} .$$

It can be regarded as a Crank-Nicolson scheme with a fixed point iteration which terminates after three iterations. It is only second order accurate in time, but has the advantage that it needs less storage and offers more temporal damping (see Rizzi et al. |239|) than the 4-stage method presented above.

Vectorization of this type of explicit methods is elementary and has been carried out for 3D aerodynamic simulation on the CYBER 205 by Rizzi |159|. All the vectors on the right-hand sides are known when needed. Evaluation of the 4-stage method leads one to the incomplete Taylor series expansion:

$$W^{(4)} = W^{(0)} - \Delta t\, PW^{(0)} + \frac{1}{2}\Delta t^2 P^2 W^{(0)} - \frac{1}{6}\Delta t^3 P^3 W^{(0)} + \frac{1}{24}\Delta t^4 P^4 W^{(0)} \quad .$$

With the aid of Horner's notation for polynomial evaluation |228| and $Q = \Delta t \cdot P$ we get:

$$W^{(4)} = \left(I - Q\left(I - Q\left(\frac{1}{2} I - Q\left(\frac{1}{6} I - \frac{1}{24} Q\right)\right)\right)\right) W^{(0)}$$

where I is the identity matrix. The somewhat modified scheme (7.1) of section 7.1 involves four equations

$$W = W^{(n)}$$

$$W = W^{(n)} - \frac{1}{4} QW$$

$$W = W^{(n)} - \frac{1}{3} QW$$

$$W = W^{n} - \frac{1}{2} QW$$

$$W^{(n+1)} = W^{(n)} - QW$$

These linked triads of the form

vector + scalar * (vector + scalar * vector)

in components, for our example,

$$W_i = WN_i + \frac{\Delta t}{4h^2} * ((W_{i-1} + W_{i+1}) - 2. * W_i)$$

increase vector performance remarkably. One further advantage is the reduction in storage requirements which is only one quarter of that of the original 4-stage scheme. In FORTRAN, the vectorized scheme, without concern for boundary conditions, may have the following simple form (with IES1 = s-1 and DTH = $\Delta t/\Delta x^2$).

```
      DO    1   K = 1,4
      F = DTH/FLOAT(5-K)
      DO    1   I = 2,IES1
    1 W(I) = WN(I)+F*(W(I-1)+W(I+1)-2.*W(I))
```

7.6 An Explicit Superstep Method for the Solution of Parabolic Differential Equations |222|

Given the linear parabolic differential equation:

$$U_t = L(U) \quad \text{in} \quad G \times (0,T)$$

with G an open sub-domain of the r-dimensional Euclidean space, E_r, $x \in E_r$, and boundary $\dot{G}$, t being the time variable, and L(V) a linear elliptic operator of the second order. We suppose that there exists an unique solution $u = u(x,t) \in G \times (0,T)$ of the differential equation, satisfying given initial and boundary conditions.

Let Δx be the spacing of an equi-distant mesh over the region and denote the vector of the approximate values of the solution u(x,t) at the mesh points at time $n \cdot \Delta t$ by the net function V^n. A well-known explicit one-step method is then:

$$V^{n+1} = V^n - \gamma L_h [V^n] \quad , \quad \gamma = \frac{\Delta t}{\Delta x^2} \; .$$

This method is known to be stable, if

$$|1 - \gamma \lambda_{max}| \leq 1$$

where λ_{max} is the largest eigenvalue of the discrete operator L_h. In the one-dimensional case, for example, with $L(U) = U_{xx}$, $\lambda_{max} = 4$, we obtain the stability condition:

$$\frac{\Delta t}{\Delta x^2} \leq \frac{1}{2} \; .$$

For r = 3 and $L(U) = U_{xx} + U_{yy} + U_{zz}$ with $\Delta x = \Delta y = \Delta z = h$ we obtain:

$$\frac{\Delta t}{h^2} \leq \frac{1}{6} \; ,$$

as a rigorous restriction on the time step Δt. However, this stability bound is in a certain sense only a sufficient one. A slight modification of the explicit method dramatically enlarges time steps. This is achieved by demanding stability not only for a single step, but for two or more time steps together. For two time steps, Δt_1, and Δt_2 for example, one obtains:

$$|(1-\gamma_1\lambda_{max})(1-\gamma_2\lambda_{max})| \leq 1$$

and with $\lambda_{max} = 4$, $r = 1$ we get:

$$\gamma_1 = \Delta t_1/\Delta x^2 = 1 - 0.5 \cdot \sqrt{2}$$

$$\gamma_2 = \Delta t_2/\Delta x^2 = 1 + 0.5 \cdot \sqrt{2} \quad .$$

It follows that:

$$\sum_{i=1}^{2} \gamma_i = 2 \qquad \text{instead of} \qquad \sum_{i=1}^{2} \gamma = 1$$

in the case where one has a known explicit method with constant time steps. The general result will be, that the presented method is n-times faster, if n is the number of the time steps belonging to one so-called "super-step". Combining n time steps to one super-step, we obtain the modified explicit one-step method:

$$V^n = \prod_{i=1}^{n} (1-\gamma_i L_h) V^0$$

with starting vector V^0 resulting from the known initial condition. Without restriction, we may transform the interval of the eigenvalues of the discrete operator into $0 < \lambda < 1$. The stability condition then is of the form:

$$|P_n(\lambda)| = |\prod_{i=1}^{n} (1-\gamma_i\lambda)| \leq 1, \; 0 \leq \lambda \leq 1 .$$

$P_n(\lambda)$ being a polynomial of degree n with zeros:

$$\lambda_i = \frac{1}{\gamma_i} \quad , \quad i = 1,2,\ldots,n .$$

The requirement:

$$\sum_{i=1}^{n} \gamma_i \stackrel{!}{=} \max$$

leads us to somewhat modified Chebyshev polynomials:

$$T_n(\lambda) = (-1)^n \cos(n \arccos(2\lambda-1))$$

(similar to those arising in Chebyshev acceleration of basic iterative methods |224|), with the zeros:

$$\lambda_i = \frac{1}{2}(\cos(\frac{2i-1}{n} \cdot \frac{\pi}{2}) + 1), \quad i = 1,2,\ldots,n .$$

If we put $P'_n(0) = T'_n(0)$, it follows that:

$$\sum_{i=1}^{n} \gamma_i = 2n^2 \quad \text{instead of} \quad \sum_{i=1}^{n} \gamma = 2n$$

of the known explicit method.

In practice, some further modifications are necessary to improve stability. Firstly, we use the zeros of certain damped Chebyshev polynomials:

$$\lambda_i = \frac{1}{2}(\cos(\frac{2i-1}{n} \cdot \frac{\pi}{2}) + 1 + \frac{R}{n^2}) \quad , \quad i = 1,2,\ldots,n$$

with damping factor $R > 0$, to enforce

$$\max_{\lambda} |P_n(\lambda)| \leq 1 .$$

Secondly we apply the zeros, not in their natural ordering but in an other permutation:

$$\tilde{\lambda}_1 = \frac{1}{\tilde{\gamma}_1} , \quad \tilde{\lambda}_2 = \frac{1}{\tilde{\gamma}_2} , \quad \ldots , \quad \tilde{\lambda}_n = \frac{1}{\tilde{\gamma}_n}$$

minimizing rounding-off errors. A measure for these is the expression:

$$\varepsilon = \sum_{K=1}^{n-1} \prod_{k=1}^{n-k} |1-\tilde{\gamma}_i|$$

which we will not discuss in further detail (see |222|).

With these improvements linear and non-linear, parabolic, differential equations and systems of them in one and two space variables, with n = 500 to n = 1000 time steps per super-step, have been solved successfully. Moreover the principle may be extended to the construction of second and third order accurate methods |222|.

A vectorization of the first-order super-step method is now straightforward, as it belongs to the class of explicit methods. Let us consider one super-step. Firstly, we compute the zeros λ_i simultaneously, permuting, and storing their reciprocals (times $\Delta t/\Delta x^2$) as the vector G. For example we put:

$$(L_h V)_j = - V_{j-1} + 2V_j - V_{j+1} \quad , \; j = 2,3,\ldots,M-1.$$

A FORTRAN version for one super-step, with N steps, may then take the following simple form:

```
      DO    1    I = 1,N
      S = G(I)
      DO    2    J = 2,M1
    2 V(J) = V(J) + S * (V(J-1) + V(J+1) - 2 * V(J))
```

with M1 = M-1, which is similar to that of section 7.5. The inner loop again represents a combination of linked triads of the form:

vector + scalar * (vector + scalar * vector) ,

which is ideal for execution on vector computers.

7.7 Finite-Difference Smoothing

In physical regions containing large gradients of the solution, amplifications of high frequency components ("wiggles") of the numerical solutions appear, depending on the fine finite-difference mesh. For example in shock waves or in stagnation regions, severe pressure gradients are to be found. In order to control these instabilities in the calculation, it proves necessary to introduce artificial viscosity into the difference scheme. There are different procedures for doing this, and many of them are easy to vectorize.

One way to obtain an artificially viscous behavior is just to select a differencing method which includes artificial viscosity implicitly due to the truncation error of the finite difference equation (see e.g. Noh and Protter |236|). A well-known example of this is the first-order up-wind differencing scheme applied to the linear, advection equation

$$U_t + a\,U_x = 0$$

which is equivalent to a second-order finite difference scheme for the equation:

$$U_t + a\,U_x = \mu U_{xx}$$

where

$$\mu = \frac{1}{2} U \Delta x \left(1 - U \frac{\Delta t}{\Delta x}\right)$$

and Δx, Δt are the grid spacings in x- resp. t-direction.

A second method is to add explicitly a diffusion term to the equations, as performed by von Neumann and Richtmyer |235| for the inviscid gas dynamic equations, in order to allow the calculation of shock waves. Then, a "shock-smearing" extends over a small number of grid intervals of the space variable. Consider now the linear, advection equation and explicitly add a diffusion term with a small "diffusion coefficient" μ. Using up-wind differencing on the advection term and forward-time centered-space differencing on the diffusion term gives

$$U_i^{n+1} = U_i^n - c(U_i^n - U_{i-1}^n) + d(U_{i+1}^n - 2U_i^n + U_{i-1}^n)$$

where $c = u\,\Delta t/\Delta x$ is the Courant number and $d = \mu \cdot \Delta t/\Delta x^2$. Here a vectorization of the additional diffusion term is straightforward since all elements involved have to be taken from the known time level n.

This method is, in principle, equivalent to the method of filtering or smoothing. The method of constructing filtering operators, such as the well-known Shuman filter, has been employed successfully to eliminate short-wave length components from numerical solutions e.g. in weather prediction (Shuman |242|) and in transonic flow (Jameson et al. |230|, Rizzi |238|, Schmidt et al. |240|). One of the simplest filters for smoothing of equally spaced data U_i is given by:

$$U_i = U_i + \gamma(U_{i-1} - 2U_i + U_{i+1})$$

with small smoothing parameters γ . The only difference between this method and the former mentioned is, that the U_i on the right-hand side of the equation have now to be taken from the new time level n+1. But these values are available because the above filter is used as a "post-process" at each time level. A simple vectorized FORTRAN loop which is easily inserted into an existing program is:

```
  DO   1      I = 2,N
1 UD(I) = U(I-1)-U(I)
  DO   2      I = 2,N1
2 U(I)   = U(I)+AMUE*(UD(I)-UD(I+1))
```

where N1 = N-1, so taking advantage of the chaining feasibilities of the CRAY. For the CYBER the DO-loop no. 2 is a linked triad of the form:

vector + scalar * vector.

In explicit descriptor notation, the FORTRAN kernel is

```
UD(2;N1)  = U (1;N1) - U (2;N1)
U (2;N2)  = U (2;N2) + AMUE * (UD(2;N2) - UD(3;N2)
```

It is reasonable to use artificial dissipation only when necessary. In computational fluid dynamics this often is achieved by setting $\gamma = \gamma_i$ (see e.g. Rizzi et al. |239|) where

$$\gamma_i = \beta_i \frac{|P_{i+1} - 2P_i + P_{i-1}|}{P_{i+1} + 2P_i + P_{i-1}}$$

is a normalized switch, which is O(1) in regions of large pressure gradients and $O(\Delta x^2)$ elsewhere. β_i is a dimensional coefficient used for the consistency of units. The resulting difference equation is second-order accurate, where the flow variable is smooth and first-order accurate elsewhere.

Again, vectorization of the filtering process with variable smoothing parameter is straightforward, since, before the smoothing process starts, the values of P_i of the pressure are all known explicitly, and the evaluation of the γ_i for i = 1,2,...,n is a vector operation.

8. References to Chapters 1 to 7

8.1 Vector and Parallel Algorithms

1. Alefeld G.: On the convergence of the symmetric SOR method for matrices with red-black ordering. Numer. Math. 39 (1982), 113-117.
2. Ames W.G.: Sparse matrix and other high performance algorithms for the CRAY-1. Systems Engineering Lab. Univ. Michigan, Report 124, 1979.
3. Azar A., Gaglot Y.: Vectorization of explicit multi-dimensional finite difference and finite element schemes. Proc. 1. Int. Coll. on Vector and Parallel Computing in Scient. Appl., Bulletin de la Direction des Etudes et Recherches, Serie C, 1 (1983), 23-29.
4. Barlow R.H., Evans D.J., Shanehchi J.: Sparse matrix vector multiplication on the ICL-DAP. Proc. Conf. on Progress in the Use of Vector and Array Processors 1983.
5. Barlow R.H., Evans D.J., Shanehchi J.: Parallel multisection for the determination of the eigenvalues of symmetric quin-diagonal matrices. Info. Processing Letters 14 (1982), 117-118.
6. Barlow R.H., Evans D.J., Shanehchi J.: Parallel multisection applied to the symmetric tridiagonal eigenvalue problem. Computer J. 25 (1982).
7. Barnes G.H., Brown R.M., Kato M., Kuck D.J.,Slotnik D.L.,Stokes R.A.: The ILLIAC IV Computer. IEEE Trans. Comp., C-17 (1968), 746-757.
8. Berendsen H.J.C., van Gunsteren W.F., Postma J.P.M.: Molecular dynamics on CRAY,CYBER and DAP. Proc.NATO Advanced Research Workshop on High-Speed Computation, Juelich, 20-22 June, 1983.
9. Berger M., Oliger J., Rodrigue G.: Predictor-corrector methods for the solution of time-dependent parabolic problems on parallel processors. In: Elliptic Problem Solvers (Schultz,M.,ed.), Acad. Press New York 1981, 197-201.
10. Bernutat-Buchmann U., Krieger J.: Analysis of elementary vector elements. In: Proc. of the 1982 Conf. on CYBER 200 in Bochum (Bernutat-Buchmann,U., Ehlich,H., Schlosser,K.-H.,eds.),Bochumer Schriften zur Parallelen Datenverarbeitung 1982, 125-142.

11. Bernutat-Buchmann U., Rudolph D., Schlosser K.-H.: Parallel Computing 1, eine Bibliographie. Bochumer Schriften zur Parallelen Datenverarbeitung (Ehlich,H.,ed.), 2. Aufl., 1983.
12. Bernutat-Buchmann U., Krieger J.: Numerische Optimierung mittels Evolutionsstrategien auf Vektorrechnern. Proc. Int. Conf. "Parallel Computing 83", North-Holland Publ. 1984.
13. Blumenfeld M.: Praekonditionierung iterativer Verfahren auf Vektorrechnern. Proc. Int. Conf. "Parallel Computing 83", North-Holland Publ. 1984.
14. Bode A., Haendler W.: Rechnerarchitektur. Springer-Verlag, Berlin 1980.
15. Book D.L.: Finite-difference techniques for vectorized fluid dynamics calculations. Springer-Verlag, New York 1981.
16. Bossavit A.: Vectorization du calcul des polynomes. Coll. AFCET "Les Math. de l'Informatique", 1982, 549-561.
17. Brandt A.: Multigrid solvers on parallel computers. In: Elliptic Problem Solvers (Schultz,M.,ed.),Acad. Press New York 1981,39-84.
18. Brent R.P.: The computational complexity of iterative methods for systems of nonlinear equations. Math. Sci. Dpmt., IBM Watson Res. Center, Yorktown Hights, New York 1972.
19. Brocard O., Bonnet C., Vigneron Y., Lejal T., Bousquet J.: A vectorized finite element method for the computation of transonic tridimensional potential flows. Proc. 1. Int. Coll. on Vector and Parallel Computing in Scient. Appl., Bulletin de la Direction des Etudes et Recherches, Serie C, 1 (1983), 45-50.
20. Brock H.K., Brooks B.J., Sullivan F.: Diamond, a sorting method for vector machines. BIT 21 (1981), 142-152.
21. Brown F.: A high performance scalar tridiagonal equation solver for the CRAY-1. Dept. Nuclear Engineering, Univ. Michigan 1980.
22. Butcher W.: The solution of the seismic one way equation on parallel computers. Proc. Int. Conf. "Parallel Computing 83", North-Holland Publ. 1984.
23. Buzbee B., Golub G., Howell J.: Vectorization for the CRAY-1 of some methods for solving elliptic difference equations. In: High Speed Computer and Algorithm Organization (Kuck,D.J.,Lawrie,D.H., Sameh,A.H.,eds.), Acad. Press, New York 1977, 255-272.
24. Buzbee B.L.: Implementing techniques for elliptic problems on vector processors. In: Elliptic Problem Solvers (Schultz,M.,ed.),

Acad. Press, New York 1981, 85-98.

25. Calahan D.A.: A vectorized general sparsity solver. Systems Engineering Lab. Univ. Michigan, Report 168, 1982.
26. Calahan D.A.: High-performance banded and profile equation solver for the CRAY-1. I. The symmetric case. Systems Engineering Lab. Univ. Michigan, Report 160, 1982.
27. Calahan D.A.: High-performance banded equation solver for the CRAY-1. II. The symmetric case. Systems Engineering Lab. Univ. Michigan, Report 166, 1982.
28. Calahan D.A.: Sparse vectorized direct solution of elliptic problems. In: Elliptic problem solvers (Schultz,M.,ed.), Acad. Press, New York 1981, 241-245.
29. Calahan D.A.: Vector processors: Models and applications. Trans. IEEE, CAS-26 (1979), 715-726.
30. Calahan D.A.: A block-oriented equation solver for the CRAY-1. Systems Engineering Lab. Univ. Michigan, Report 136, 1980.
31. Calahan D.A.: Vectorized sparse elimination. Scient. Comp. Info. Exchange Meeting, Livermore, Cal. Sept. 1979, 103-114.
32. Calahan D.A., Ames W.G., Sesek E.J.: A collection of equation solving codes for the CRAY-1. Systems Engineering Lab. Univ. Michigan, Report No. 133, 1979.
33. Calahan D.A.: Vectorized sparse equation solution on the CRAY-1. Proc. 1979 Int. Conf. Parallel Computing, Wayne State Univ., 1979.
34. Chen S.C., Kuck D.J., Sameh A.H.: Practical parallel band triangular system solvers. ACM Trans. Math. Software 4 (1978), 270-277.
35. Chima R.V., Johnson G.M.: Efficient solution of the Euler and Navier-Stokes equations with a vectorized multiple-grid algorithm. AIAA-Paper 83-1893 (1983).
36. CRAY-1 FORTRAN Reference Manual (CFT). Pub. 2240009, CRAY Research, Minneapolis 1979.
37. CRAY Research Inc. Optimization guide. CRAY Res. Pub. SN-0220, 1981.
38. CYBER 205 Service:Efficient FORTRAN techniques.Control Data User Guide 84002680, 1982.
39. CYBER 205 Service: VAST automatyic vectorizor. Control Data User Guide 84002690, 1982.
40. Diekkaemper R.: Vektorrechner orientierte Finite-Element-Analyse

bei nichtlinearen Problemen in der Strukturmechanik. Proc. Int. Conf. "Parallel Computing 83", North-Holland Publ. 1984.

41. Dongarra J.J.: Redesigning linear algebra algorithms. Proc. 1. Int. Coll. on Vector and Parallel Computing in Scient. Appl., Bulletin de la Direction des Etudes et Recherches, Serie C, 1 (1983), 51-59.

42. Dongarra J.J., Hinds A.R.: Unrolling loops in Fortran. Software-Practice and Experience 9 (1979), 219-229.

43. Dubois P., Rodrigue G.: An analysis of the recursive doubling algorithm. Lawrence Livermore Lab., UCRL-79071, 1971.

44. Dubois P.F., Rodrigue G.H.: Operator splitting on the STAR without transposing. Lawrence Livermore Lab., UCID-17515, 1977.

45. Dubois P.F., Rodrigue G.H.: Operator splitting on vector processors. Lawrence Livermore Lab., UCRL-79316, 1977.

46. Dubois P.F., Greenbaum A., Rodrigue G.: Approximating the inverse of a matrix for use in iterative algorithms on vector processors. Computing 22 (1979), 257-268.

47. Duff I.S.: The solution of sparse linear equations on the CRAY-1. Proc.NATO Advanced Research Workshop on High-Speed Computation Juelich, 20-22 June, 1983.

48. Engeln-Muellges G., Sommer A.: Bibliographie: Parallel-Rechner, Parallele Algorithmen, Parallele Programmierung. Report, Computer Center, RWTH Aachen 1980.

49. Engeln-Muellges G.:Systemuntersuchung zur Vektormaschine CRAY-1. Report des Rechenzentrums der RWTH Aachen, Aug. 1980.

50. Erhel J., Lichnewsky A., Thomasset F.: Parallelism in finite element computation. Proc. IBM Symp. on Vector Computers and Sc. Comp., Rome 1982.

51. Evans D.J., Hadjidimos A., Noutsos D.: The parallel solution of banded linear equations by the new quadrant interlocking factorization method. Int. J. Comp. Math. 9 (1981), 151-162.

52. Evans D.J.: The parallel solution of partial differential equations. Proc. Int. Conf. "Parallel Computing 83", North-Holland Publ. 1984.

53. Feilmeier M. (ed.): Parallel computers - parallel mathematics. North-Holland, Amsterdam 1977.

54. Feilmeier M., Joubert G., Segerer G., Schendel U.: Parallele Datenverarbeitung und parallele Algorithmen. TU Berlin, Brennpunkt

Kybernetik 1979.

55. Feilmeier M., Roensch W.: Parallel nonlinear algorithms. Computer Physics Comm. 26 (1982), 335-348.

56. Feilmeier M.: Parallel numerical algorithms. In: Parallel Processing Systems (Evans,D.J.,ed.), Cambridge Univ. Press 1982.

57. Feilmeier M., Joubert G., Schendel U. (eds.): Proceedings of "Parallel Computing 83", North-Holland Publ. 1984.

58. Flanders P.M., Hunt D.J., Reddaway S.F., Parkinson D.: Efficient high speed computation with the DAP. In: High Speed Computer and Algorithm Organization (Kuck,D.J.,Lawrie,D.H.,Sameh,A.H.,eds.), Acad.Press 1977, 113-128.

59. Flanders P.M.: Fortran extensions for a highly parallel processor. ICL 1979.

60. Flynn M.J.: Some computer organizations and their effectiveness. IEEE Trans. Comput. C-21 (1972), 948-960.

61. Fong K., Jordan T.L.: Some linear algebraic algorithms and their performance on the CRAY-1. Los Alamos Scient. Lab., Report LA-6774, 1977.

62. Fornberg B.: A vector implementation of the Fast Fourier Transform algorithm. Math. Comp. 36 (1981), 189-191.

63. Garry J.M.: Analysis of application programs and software requirements for high speed computers.In:High Speed Computer and Algorithm Organization (Kuck,D.J., Lawrie,D.H., Sameh,A.H.,eds.), acad.Press, New York 1977, 329-354.

64. Gentzsch W., Mueller-Wichards D., Weiland C.: Moeglichkeiten und Probleme bei der Anwendung von Vektorrechnern. DFVLR IB 221-81 A 05 (1981).

65. Gentzsch W.: High performance processing needs in fluid dynamics. Proc. SEAS Spring Meeting, Amsterdam 1982, 575-590.

66. Gentzsch W.: A survey of the new vector computers CRAY-1S, CDC-CYBER 205 and the parallel computer ICL-DAP. Architecture and programming (in german). DFVLR-FB 82-02 Report, Koeln 1982.

67. Gentzsch W.: How to maintain the efficiency of highly serial algorithms involving recursions on vector computers. Proc. 1. Int. Coll. on Vector and Parallel Computing in Scient. Appl.,Bulletin de la Direction des Etudes et Recherches, Serie C, 1 (1983), 79-86.

68. Gentzsch W.: Benchmark results on physical flow problems. Proc.

Conf. High-Speed Comput., Juelich 1983.
69. Gentzsch W., Schaefer G.: Solution of large linear systems on vector computers. Proc. Int. Conf. "Parallel Computing 83", North-Holland Publ. 1984.
70. George A., Poole W.G., Voigt R.G.: Analysis of dissection algorithms for vector computers. ICASE, Hampton, VA, 1976.
71. Gietl H.: Anpassung zweier numerischer Algorithmen aus dem Bereich der partiellen Differentialgleichungen zur effizienten Verwendung auf einem Vektorrechner. Proc. Int. Conf. "Parallel Computing 83", North-Holland Publ. 1984.
72. Giloi W.K.: Rechnerarchitektur. Springer-Verlag, Berlin 1981.
73. Giroux E.D.: Vectorization of a complex physics calculation utilizing techniques suitable for many types of array processors. Proc. Int. Conf. "Parallel Computing 83", North-Holland Publ. 1984.
74. Gostick R.W.: Software and algorithms for the Distributed-Array Processors. ICL Techn. J. 1979, 116-135.
75. Greenbaum A.,Rodrigue G.:The ICCG method for the STAR. Res. REP. UCID-17574, Lawrence Livermore Lab., 1977.
76. Groginsky H.L., Works G.A.: A pipeline Fast Fourier Transform. IEEE Trans. Comp., C-19 (1970), 1015-1019.
77. Haendler W.: Feldrechner. Informatik Spektrum 2 (1979), 165-166.
78. Haendler W. (ed.): Proc. CONPAR 81. Lecture Notes in Computer Science, Vol. 111, Springer Berlin 1981.
79. Hankey W.L., Graham J.E., Shang J.S.: Navier-Stokes solution of a slender body of revolution at large incidence. AIAA-Paper 81-0190 (1981).
80. Heller D.E.: Some aspects of the cyclic reduction algorithm for block tridiagonal linear systems. SIAM J. Numer. Anal., 13 (1976), 484-496.
81. Heller D.E., Stevenson D.K., Traub J.F.: Accelerated iterative methods for the solution of tridiagonal systems on parallel computers. JACM, 23 (1976), 636-654.
82. Heller D.: A survey of parallel algorithms in numerical linear algebra. SIAM Review 20 (1978), 740-777.
83. Higbie L.: Vectorization and conversion of FORTRAN programs for the CRAY-1 (CFT) compiler. Pub. 2240207, CRAY Research, Minneapolis 1979.

84. Higbie L.: Applications of vector processing. Computer Design, April 1978.

85. Hockney R.W., Jesshope C.R.: Parallel Computers - Architecture, Programming and Algorithms. Adam Hilger, Bristol 1981.

86. Hockney R.W.: Performance of parallel computers. Proc. NATO Advanced Research Workshop on High-Speed Computation, Juelich, 20-22 June, 1983.

87. Hodous M.F., Bozek D.G., Ciarelli D.M., Ciarelli K.J., Kline K.A., Katnik R.B.: Vector processing applied to boundary element algorithms on the CDC CYBER 205. Proc. 1. Int. Coll. on Vector and Parallel Computing in Scient. Appl., Bulletin de la Direction des Etudes et Recherches, Serie C, 1 (1983), 87-94.

88. Hord R.M.: The ILLIAC IV. Springer-Verlag, New York 1982.

89. Hossfeld F.: Parallelverarbeitung - Konzepte und Perspektiven. Angewandte Informatik, 22 (1980), 485-492.

90. Hossfeld F.: Parallelprozessoren und Algorithmenstruktur. Report No. 87, KFA Juelich 1980.

91. Hossfeld F.: Parallele Algorithmen. Reihe Informatik-Fachberichte, Bd. 64, Springer Verlag, Berlin 1983.

92. Hossfeld F., Weidner P.: Parallele Algorithmen. Informatik-Spektrum 6 (1983), 142-154.

93. Hunt D.J.: Application techniques for parallel hardware. ICL 1979.

94. Hunt D.J.: Numerical solution of Poisson's equation on an array processor using iterative techniques. DAP Techn. Report No. 1, 1981.

95. Hunt D.J.: A study of finite element analysis on DAP. DAP Techn. Report No. 2, 1981.

96. Johnson O.G., Paul G.: Vector algorithms for elliptic partial differential equations based on the Jacobi method. In: Elliptic Problem Solvers(Schultz,M.,ed.),Acad.Press New York 1981, 345-351.

97. Johnson P.M.: CRAY-1 computer system. Publ. No. 2240002A, CRAY Research, Minneapolis 1977.

98. Johnson P.M.: An introduction to vector processing. Computer Design 1978, 89-97.

99. Jordan H.F.: Standard synchronization in HEP Fortran. Denelcor Inc. Publication 1982.

100. Jordan H.F.: Performance measurements on HEP - a pipelined MIMD computer. Univ. of Colorado, Boulder Col. 80309, Dec. 1981.

101. Jordan H.F.: Parallel programming on the HEP multiple instruction stream computer. Denelcor, Jan. 8, 1982.

102. Jordan T.L.: A new parallel algorithm for diagonally dominant tridiagonal matrices. Los Alamos Scientific Lab. Report 1974.

103. Joubert R.G.: Parallel computer report. Computer Science Dep., Univ. of Natal, Durban 1977.

104. Kai Hwang, Shun-Piao Su: Vector computer architecture and processing techniques. Advances in Computers (1981), 115-197.

105. Kascic M.J.: Lecture Notes 2+3 of vector class. CDC Minneapolis 1978.

106. Kascic M.J.: A direct poisson solver on STAR. CDC Minneapolis 1978.

107. Kascic M.J.: Vector processing on the Cyber 200. Infotech State of the Art Report "Supercomputers", Infotech Int. Ltd., Maidenhead, U.K. 1979.

108. Kascic M.J.: Vector processing on the CYBER 200. Angewandte Informatik 22 (1980), 27-37.

109. Kascic M.J.: Anatomy of a Poisson solver. Proc. Int. Conf. "Parallel Computing 83", North-Holland Publ. 1984.

110. Keller J.D., Jameson A.: Preliminary study of the use of the STAR-100 computer for transonic flow calculations. AIAA-Paper 78-12 (1978).

111. Kershaw D.S.: The solution of single linear tridiagonal systems and vectorization of the ICCG algorithm on the CRAY-1. Res. Rep. UCID - 19085, Lawrence Livermore Lab., 1981.

112. Kirchheimer F.: Zwei vektorisierte Routinen aus einem geophysikalischen Anwendungsprogramm. Proc. Int. Conf. "Parallel Computing 83", North-Holland Publ. 1984.

113. Kogge P.M.: The numerical stability of parallel algorithms for solving recurrence problems. Digital Systems Lab., Stanford Univ. 1972.

114. Kogge P.M.: Maximal rate pipeline solutions to recurrence problems. Proc. 1. Ann. Symp. on Comp. Architectures (1973), 71-76.

115. Kogge P.M.: A parallel algorithm for the efficient solution of a general class of recurrence equations. IEEE Trans. Comp., C-22 (1973), 786-793.

116. Kogge P.M.: Algorithm development for pipeline processors. Proc.

1. Ann. Conf. on Computer Arch. 1973, 71-80.
117. Kogge P.M.: Parallel solution of recurrence problems. IBM J.Res. Dev., 18 (1974), 138-148.
118. Kogge P.M.: The architecture of pipelined computers. Acad. Press, New York 1981.
119. Korn D.G., Lambiotte J.J.: Computing the Fast Fourier Transform on a vector computer. Math. Comp. 33 (1979), 977-992.
120. Kozdrowicki E.W., Theis D.J.: Second generation of vector supercomputers. IEEE Comput. Mag. 13 (1980), 71-83.
121. Kuck D.J.: ILLIAC IV software and application programming. IEEE Trans. Comp., C-17 (1968), 758-770.
122. Kuck D.J., Lawrie D.H., Sameh A.H.: High speed computer and algorithm organization. Acad. Press, New York 1977.
123. Kuck D.J.: The structure of computers and computations. Wiley New York 1978.
124. Kuck D.J.: Automatic program restructuring for high-speed computation. Proc. CONPAR 81 Conf. on Analysing Problem - Classes and Programming for Parallel Computing, Nuernberg F.R.G. 1981.
125. Lambiotte J.J., Voigt R.G.: The solution of tridiagonal linear systems on the CDC STAR-100 computer. ACM Transactions on Math. Software, 1 (1975), 308-329.
126. Lambiotte J.J.: The solution of linear systems of equations on a vector computer. PH.D. Thesis, Univ. Virginia 1975.
127. Lichnewsky A.: Sur la resolution de systemes lineaires issus de la methode des elements finis par une machine multiprocesseurs. Rapp. Rech. 119, INRIA 1981.
128. Lichnewsky A.: Some vector and parallel implementation for preconditioned conjugate gradient algorithms. Proc. NATO Advanced Research Workshop on High-Speed Computation, Juelich, 20-22 June, 1983.
129. Lomax H.: Some prospects for the future of computational fluid dynamics. AIAA-Paper 81-0994, 1981.
130. MacCormack R.W., Stevens K.G.: Fluid dynamics applications of the ILLIAC IV computer. In: Computational Methods and Problems in aeronautical fluid dynamics (Hewitt,ed.), Acad.Press, New York 1976, 448-465.
131. Madson N.K., Rodrigue G.H.: A comparison of direct methods for tridiagonal systems on the CDC-STAR 100. Lawrence Livermore Lab.,

UCRL - 76993, 1976.

132. Madson N.K., Rodrigue G.H., Karush J.I.: Matrix multiplication by diagonals on a vector/parallel processor. Info Processing Letter, 5 (1976).

133. Matsuno K.: A vector-oriented finite difference scheme for calculating 3-D compressible laminar and turbulent boundary layers on practical wing configurations. AIAA-Paper 81-1020 (1981).

134. Meurant, G.: Vector preconditionning for the conjugate gradient method. To appear 1983.

135. Modi J.J., Parkinson D.: Study of Jacobi methods for eigenvalues and singular value decomposition on DAP. Computer Physics Communications 26 (1982), 317-320.

136. Morawe M.: Parallelverarbeitung in der Seismik. Proc. Int. Conf. "Parallel Computing 83", North-Holland Publ. 1984.

137. Morjaria M., Makinson G.J.: Operations with sparse matrices on the ICL-DAP. Math. Inst. Univ. Kent, Canterbury 1982.

138. Mossberg B.: An informal approach to number crunching on the CYBER 205. Control Data Corp., Pub. No. 84002390.

139. Mossberg B.: Sorting on the CYBER 205. Control Data Corp., June 1982.

140. Mueller-Wichards D., Gentzsch W.: Performance comparisons among several parallel and vector computers on a set of fluid flow problems. DFVLR IB 262-82 R 01 Report, 1982.

141. Nolen J.S., Kuba D.W., Kascic M.J.: Application of vector processors to the solution of finite difference equations. Soc. Petr. Eng. AIME 1979.

142. Oed W., Lange O.: Transforming linear recurrence relations for vector processors. Proc. Int. Conf. "Parallel Computing 83", North-Holland Publ. 1984.

143. Ortega J.M.,Voigt R.G.: Solution of partial differential equations on vector computers. ICASE Report 77-7, NASA Langley 1977.

144. Parkinson D.: An introduction to array processors. Systems Int., Nov. 1977.

145. Parkinson D., Flanders P.M.: Fourier transformation on DAP. DAP Technical Report No. 7, 1981.

146. Parkinson D.: The solution of sets of equations. DAP Technical Report No. 18, 1981.

147. Parkinson D., Liddell H.M.: The measurement of performance on a

high parrallel system. IEEE Trans. Comput. C-32 (1983), 32-37.
148. Parkinson D.: Experience in exploiting large scale parallelism. Proc.NATO Advanced Research Workshop on High-Speed Computation Juelich, 20-22 June, 1983.
149. Paul G., Wilson M.W.: The VECTRAN Language: An experimental language for vector/matrix array processing. IBM Palo Alto Scientific Center Report 6320-3334, 1975.
150. Petersen W.P.: Basic linear algebraic subprograms for CFT usage. Pub. 2240208, CRAY Research, Minneapolis 1979.
151. Petersen W.P.: Linear digital filters for CFT usage. CRAY Research, Technical Note 2240210, 1979.
152. Pulliam T.H., Lomax H.: Simulation of 3-D compressible viscous flow on the ILLIAC IV computer. AIAA-Paper 79-0206 (1979).
153. Ramamoorthy C.V., Li H.F.: Pipeline architecture. Computing Surveys, 9 (1977), 61-102.
154. Reddaway S.F.: The DAP approach. ICL 1979.
155. Reddaway S.F., Hunt D.J.: Study of a finite element problem. DAP Technical Report No. 10, 1981.
156. Redhed D.D., Chen A.W.: New approach to the 3-D transonic flow analysis using the STAR-100 computer. AIAA-Journal 17 (1979), 98-99.
157. Regionales Rechenzentrum, Universitaet Hannover: Neue Rechnerarchitekturen, Berichte 1980/82.
158. Remund R.N., Taggart K.A.: "To vectorize" or to "vectorize": That is the question. In: High Speed Computer and Algorithm Organization (Kuck,D.J.,Lawrie,D.H.,Sameh,A.H.,eds.),Acad.Press New York 1977, 399-410.
159. Rizzi A.: Vector coding the finite volume procedure for the CYBER 205. VKI-Lecture Series 1983-04.
160. Rodrigue G.H., Madson N., Karush J.: Odd-even reduction for banded linear equations. Lawrence Livermore Lab., UCRL-78652, 1976.
161. Rodrigue G.H.: Operator splitting on the STAR without transposing. Lawrence Livermore Lab., UCRL-17515, 1977.
162. Rodrigue G.: Parallel Computations. Acad.Press, New York 1982.
163. Roensch W.: Stabilitaets- und Zeituntersuchungen arithmetischer Ausdruecke auf dem Vectorrechner CRAY-1S. Dissertation, TU Braunschweig 1983.

164. Roensch W.:Timing and stability analysis of arithmetic expressions on the CRAY-1. Proc. Int. Conf. "Parallel Computing 83", North-Holland Publ. 1984.
165. Rudsinki L.,Worlton J.: The impact of scalar performance on vector and parallel processors. In: High Speed Computer and Algorithm Organization (Kuck,D.J.,Lawrie,D.H.,Sameh,A.H.,eds.),Acad.Press, New York 1977, 451-452.
166. Rudsinki L., Pieper G.W.: Evaluating computer program performance on the CRAY-1. Argonne Nat. Lab. Report ANL-79-9, 1979.
167. Russel R.M.: The CRAY-1 computer system. CACM 21 (1978), 63-72.
168. Sack R.A.: Relative pivoting for systems of linear equations. Proc. Int. Conf. "Parallel Computing 83", North-Holland Publ. 1984.
169. Sameh A.H., Chen S.C., Kuck D.J.: Parallel Poisson and biharmonic solvers. Computing 17 (1976), 219-230.
170. Sameh A.H., Brent R.P.: Solving triangular systems on a parallel computer. SIAM J. Numer. Anal. 14 (1977), 1101-1113.
171. Sameh A.: Numerical parallel algorithms - A survey. In: High Speed Computer and Algorithm Organization (Kuck,D.J., Lawrie,D.H., Sameh,A.H.,eds.), Acad. Press, New York 1977, 207-228.
172. Schendel U.: Parallele Numerik. Lecture Notes, FU Berlin 1977.
173. Schendel U.,Brandenburger J.:Algorithmen zur Loesung rekurrenter Relation. Preprint 101/79, FU Berlin 1979.
174. Schendel U.: Einfuehrung in die parallele Numerik. Oldenburg Verl. Muenchen 1981.
175. Scarborough R.G.: Writing optimizable Fortran. IBM Scientific Center, Palo Alto, Ca.
176. Schmidt W.: Numerische Simulation - eine Schluessel-Technologie nicht nur fuer die Luftfahrt. CRAYationen 1983, 1-7.
177. Schnepf E., Schoenauer W.: Parallelization of PDE software for vector computers. Proc. Int. Conf. "Parallel Computing 83", North-Holland Publ. 1984.
178. Schoenauer W., Schnepf E.,Raith K.: The redesign and vectorization of the SLDGL - program package for the selfadaptive solution of nonlinear systems of elliptic and parabolic PDE's.IF/P Working Group 2.5 on Numer. Software. PDE-Software: Modules, Interfaces and Systems, Soderkoping, Sweden, Aug. 22-26, 1983.
179. Schoenauer W., Raith K.: A polyalgorithm with diagonal storing for

the solution of very large indefinite linear banded systems on a vector computer. Proc. 10th IMACS World Congress on System Simulation and Scientific Computation 1 (1982), 326-328.

180. Schoenauer W.: Vektorrechner: Die schnellsten Rechenanlagen. Jahrbuch Ueberblicke Mathematik, Bibl. Inst. Mannheim 1982.

181. Schoenauer W.: The efficient solution of large linear systems, resulting from the FDM for 3-D PDE's, on vector computers. Proc. 1. Int. Coll. on Vector and Parallel Computing in Scient. Appl., Bulletin de la Direction des Etudes et Recherches, Serie C, 1 (1983), 135-142.

182. Schoenauer W.: Numerical experiments with instationary Jacobi-OR methods for the iterative solution of linear equations. ZAMM 63 (1983), T380-T382.

183. Schreiber R., Wei-Pei Tang: Vectorizing the conjugate gradient method. To appear 1983.

184. Shang J.S., Buning P.G., Hankey W.L., Wirth M.C., Calahan D.A., Ames W.: Numerical solution of the 3-D Navier-Stokes equations on the CRAY-1 computer. Proc. Scientific Comp. Inf. Exchange Meeting (1979), 159-166.

185. Shang J.S., Buning P.G., Hankey W.L., Wirth M.C.:Performance of a vectorized three-dimensional Navier-Stokes code on the CRAY-1 computer. AIAA-Journal 18 (1980), 1073-1078.

186. Shang J.S.: Numerical simulation of wing-fuselage interference. AIAA-Paper 81-0048 (1981).

187. Simpson R.B., Yazici A.: An organization of the extrapolation method for vector processing. Proc. Int. Conf. "Parallel Computing 83", North-Holland Publ. 1984.

188. Sintz R.H.: Optimal use of a vector processor. Digest of Papers, IEEE COMPCON (1980), 277-281.

189. Smith B.J.: The evolution of a supercomputer. Denelcor Inc. Publication.

190. Smith B.J.: A pipelined, shared resource MIMD computer. Proc. Int. Conf. Parallel Processing 1978, 6-8.

191. Smith B.J.: Architectur and applications of the HEP multiprocessor computer system. Real Time Signal Processing IV, Proc. of SPIE, Vol. 298 (1980).

192. Smith R.E., Pitts J.I.,Lambiotte J.J.:A vectorization of the Jameson-Caughey transonic swept-wing computer program FLO-22 for

the STAR-100 computer.NASA Techn.Memorandum TM-78665 (1978).
193. South J.C., Keller J.D., Hafez M.M.: Vector processor algorithms for transonic flow calculations. AIAA-Journal 18 (1980), 786-792.
194. Spradley L.W., Stalnaker J.F., Ratliff A.W.: Solution of the three-dimensional Navier-Stokes equations on a vector processor. AIAA-Journal 19 (1981), 1302-1308.
195. Spradley L.W., Stalnaker J.F., Ratliff A.W.: Hyperbolic/parabolic development for the GIM-STAR code. NASA Contractor Report 3369 (1980).
196. Stone H.S.: Problems of parallel computation. In: Complexity of Sequential and Parallel numerical algorithms (Traub,J.F.,ed.), Acad.Press, New York 1973, 1-16.
197. Stone H.S.: An efficient parallel algorithm for the solution of a tridiagonal linear system of equations. JACM, 20 (1973), 27-38.
198. Stone H.S.: Parallel tridiagonal solvers. Digital Systems Lab., Stanford Univ. 1974. And: J. ACM 20 (1973), 27-38.
199. Stone H.S.: Sorting on STAR. IEEE Trans. Software Eng., SE-4 (1978), 138-146.
200. Swarztrauber P.N.: Vectorizing the FFT's. In:Parallel Computations (Rodrigue,G.,ed.), Acad.Press, New York 1982.
201. Temperton C.:Fast Fourier Transforms on the CYBER 205. Proc. NATO Advanced Research Workshop on High-Speed Computation, Juelich, 20-22 June, 1983.
202. Temperton C.:Fast Fourier Transform and Poisson solvers on CRAY-1 In: Supercomputers, Infotech State of the Art Report, Maidenhead, U.K., 1979.
203. Temperton C.: Fast Fourier Transform for numerical prediction models on vector computers. Proc. 1. Int. Coll. on Vector and Parallel Computing in Scient. Appl., Bulletin de la Direction des Etudes et Recherches, Serie C, 1(1983), 159-162.
204. Traub J.F.: Iterative solution of tridiagonal systems on parallel or vector computers. In: Complexity of Sequential and Parallel Numerical Algorithms (Traub,J.F.,ed.), Acad. Press New York 1973.
205. Traub J.F. (ed.): Complexity of sequential and parallel numerical algorithms. Acad. Press New York 1977.
206. Tuttle P.G.: Implementation of selected eigenvalue algorithms on a vector computer. M.S. thesis, Univ.Virginia, Charlottesville 1975.
207. van der Vorst H.A.: On the vectorization of some simple ICCG

methods. Proc. 1. Int. Coll. on Vector and Parallel Computing in Scient. Appl., Bulletin de la Direction des Etudes et Recherches, Serie C, 1 (1983).

208. van der Vorst H.A.: A vectorizable variant of some ICCG methods. SIAM J. Sci. Stat. Comput. 3 (1982), 350-356.

209. Verpat M.: Mise en oeuvre et performances d'algorithmes sur un calculateur vectoriel. These Univ. Paul Sabatier, Toulouse 1982.

210. Verprat M., Thomas J.M.: Implementation of some algorithms and their performance on a vector computer. Proc. 1. Int. Coll. on Vector and Parallel Computing in Scient. Appl., Bulletin de la Direction des Etudes et Recherches, Serie C, 1 (1983), 163-166.

211. Voigt R.G.: The influence of vector computer architecture on numerical algorithms. ICASE Report 77-8, NASA Langley 1977.

212. Wang H.H.: On vectorizing the Fast Fourier Transform. BIT 20 (1980), 233-2433.

213. Ward R.C.:The QR-algorithm and Hymans method on vector computers. Math. Comp. 30 (1976), 132-142.

214. Wiese G.: Entwicklung, Laufzeitanalyse und Optimierung von CRAY-Assembler-Programmen. Diplomarbeit TU Braunschweig, F.R.G. 1983.

215. Zabolitzky J.G.: Vector programming of Monte-Carlo and numerical problems. In: Proc. of the 1982 Conf. on CYBER 200 in Bochum (Bernutat-Buchmann,U., Ehlich,H., Schlosser,K.-H.,eds.), Bochumer Schriften zur Parallelen Datenverarbeitung 1982,165-174.

8.2 Further Applications

216. Abou EL-Seoud S., Toernig W.: Solution of large sparse nonlinear systems by monotone convergent iterations and applications. ZAMM 63 (1983), 103-113.

217. Beam R.M., Warming R.F.: An implicit factored scheme for the compressible Navier-Stokes equations. AIAA-Journal 18 (1980), 393-402.

218. Beam R.M., Warming R.F.: Implicit numerical methods for the compressible Navier-Stokes and Euler equations. VKI-Lecture Series, Bruessel 1982-04.

219. Buzbee B.L., Golub G.H., Nielson C.W.: On direct methods for solving Poisson's equations. SIAM J. Numer. Anal. 7 (1970), 617-656.

220. Chodura R., Schlueter A.: In search of stable 3D MHD-equilibria. Proc. 2. Europ. Conf. Comp. Phys., Garching 1976.

221. Engeli M., Ginsburg M., Rutishauser H., Stiefel E.: Refined iterative methods for the computation of the solution and the eigenvalues of self-adjoint boundary value problems. Mitt. Inst. Angew. Math. ETH, Zuerich, Nr. 8, 1959.

222. Gentzsch W.: Numerical solution of linear and non-linear parabolic differential equations by a time-discretization of third order accuracy. Proc. 3. GAMM Conf. Numer. Meth. in Fluid Mechanics. Notes on Numerical Fluid Mechanics, Vol.2 (Hirschel, E.H.,ed.), Vieweg-Verlag, Braunschweig 1980, 109-117.

223. Gustavson F.G.: Some basic techniques for solving sparse systems of linear equations. In: Sparse Matrices and Their Application (Rose, ed.), Plenum Press 1972, 41-52.

224. Hageman L.A., Young D.M.: Applied iterative methods. Acad.Press, New York 1981.

225. Hirschel E.H., Kordulla W.: Shear flow in surface-oriented coordinates. Notes on Numerical Fluid Mechanics, Vol.4, Vieweg-Verlag, Braunschweig 1981.

226. Hockney R.W.: A fast direct solution of Poisson's equation using Fourier analysis. J. Assoc. Comput. Mach. 12 (1965), 95-113.

227. Holt M.: Numerical methods in fluid dynamics. Springer-Verlag, New York 1977.

228. Isaacson E., Keller H.B.: Analysis of numerical methods. Wiley and Sons, New York 1966.

229. Jameson A., Caughey D.A.: Numerical calculation of the transonic flow past a swept wing. ERDA Res. and Dev. Rept. C00-3077-140, New York Univ. 1977.

230. Jameson A., Schmidt W., Turkel E.: Numerical solutions of the Euler equation by finite volume methods using Runge-Kutta time-stepping schemes. AIAA-paper no. 81-1259, Palo Alto 1981.

231. Kraus W.: Panel methods in aerodynamics. Numerical methods in fluid dynamics (Wirz, H.J., Smolderen J.J.,eds.), Hemisphere Publ. Corporation, Washington 1978, 237-297.

232. Larson J., Sameh A.: Efficient calculation of the effects of

rounding errors. ACM Trans. Math. Softw. 4 (1978), 228-236.
233. Metcalf M.: Fortran Optimization. Acad.Press, New York 1982.
234. Miles G.A., Stewart K.L., Tee G.T.: Elementary divisors of the Liebmann Process. Comput. J. 6 (1964), 353-355.
235. Neumann J. von, Richtmyer R.D.: A method for the numerical calculations of hydrodynamical shocks. J. Appl. Phys. 21 (1950), pp. 232.
236. Noh W.F., Protter M.H.: Difference methods and the equations of hydrodynamics. J. Math. Mech. 12 (1963), pp. 149.
237. Ortega J.M., Rheinboldt W.C.: Iterative solution of nonlinear equations in several variables. Acad. Press, New York 1970.
238. Rizzi A.: Computation of rotational transonic flow. In: Numerical Methods for the Computation of Inviscid Transonic Flows with Shock Waves. Notes on Numerical Fluid Mechanics, Vol.3 (Rizzi,A., Viviand,H.,eds.), Vieweg Verlag, Braunschweig 1981.
239. Rizzi A., Eriksson L.E.: Explicit multistage finite volume procedure to solve the Euler equations for transonic flow. VKI-Lecture Series 1983-4.
240. Schmidt W., Jameson A.: Recent developments in finite-volume time-dependent techniques for two- and three-dimensional transonic flows. VKI-Lecture Series 1982-04.
241. Schwamborn D.: Boundary layers on wings. Proc. 5. GAMM-Conf. on Numer. Meth. in Fluid Mechanics, Rome 1983. Notes on Numerical Fluid Mechanics, Vol.7 (Pandolfi M., Piva R., eds.), Vieweg-Verlag, Braunschweig 1984.
242. Shuman F.G.: Numerical methods in weather prediction: II. Smoothing and filtering. Monthley Weather Rev. 85 (1957), 357-361.
243. Stummel F.: Perturbation theory for evaluation algorithms of arithmetic expressions. Math. Comp. 37 (1981), 435-473.
244. Toernig W.: Monoton einschliessend konvergente Iterationsprozesse vom Gauss-Seidel-Typ zur Loesung nichtlinearer Gleichungssysteme im R^N und Anwendungen. Math. Meth. Appl. Sci. 2 (1980), 489-503.

9. Vectorization of Algorithms in Computational Fluid Dynamics on the CRAY-1 Vector Computer

9.1 MacCormack's Methods and Vectorization

Willi Kordulla, DFVLR

9.1.1 Introduction

Among computational fluid dynamicists the tremendous impact of MacCormack's contributions to the development of computational tools for integrating the governing equations for inviscid and viscous flows is well recognized, and this is documented by numerous publications based on the use of his methods for integrating steady as well as time-dependent Euler and Navier-Stokes equations. Best known are MacCormack's purely explicit predictor-corrector versions |1,2| of the two-level scheme of Lax and Wendroff. Explicit schemes are easily applied, and can, in general, be completely vectorized, but they have to satisfy severe stability conditions with respect to the marching step size. Such conditions are most stringent for viscous flows if the wall-normal, boundary-layer direction is to be resolved properly, with step sizes for turbulent flows of the order of 0.00005 times the characteristic length involved. For the explicit method to be stable in time-dependent calculations, the time-wise step must then be chosen proportional to the square of the corresponding spatial step size divided by the kinematic viscosity. In the case of inviscid flow described by the Euler equations, the time step size has to be proportional only to the spatial step size itself divided by an appropriate velocity, this being the Courant-Friedrichs-Lewy condition. To overcome these restrictions, at least partially, MacCormack introduced symmetric operator splitting (i.e. the concept of breaking up the multi-dimensional operator into sequences of one-dimensional operators without factorization) such that one large step is performed for the surface tangential direction, while for the crucial wall-normal direction that time step is divided into many single steps so that all steps satisfy the stability conditions |2,5,6|. While the Euler equations in three dimensions |4,5| were integrated successfully and in CPU times which were reasonable for serial computers, this could not be done for the Navier-Stokes equations. Even in two dimensions , in particular for high-

Reynolds number transonic flows |6|, the solutions were very time consuming.

In the mid-seventies, the development of fully implicit schemes was favoured for integrating the time dependent, governing equations for aerodynamic flows, because, according to the stability analysis of linear model equations, such methods are not restricted to small time steps. In practice restrictions do occur due to the method of implementing the boundary conditions, due to the non-linearity of the equations, or, of course, just due to the necessity of resolving physical features of the flow in question. In the case where the small steps are not imposed by the physical considerations, but instead by stability requirements, additional numerical diffusion terms are usually added to enhance the capability of taking larger steps; e.g. in a time dependent calculation where only the steady state is of interest. To be more competitive with such schemes, MacCormack modified his method in such a way that the explicit scheme is taken only for the surface tangential direction, while a hybrid characteristics-implicit method is being used with respect to the wall-normal direction |7|. For the latter, the equations were split into the convective Euler and into the diffusion part so that a method of characteristics and an implicit scheme, e.g. the trapezoidal rule, could be used accordingly. For large Reynolds number flows the new method was shown to be about two orders of magnitude faster than the former explicit scheme, and solutions for three-dimensional, turbulent, supersonic flows in reasonable computer times were feasible using MacCormack's method |8|. However, considering the advent of vector computers, the hybrid scheme was not well suited to the requirements for optimum performance of such machines due to the method of characteristics involved [9]. Note, that fully implicit schemes are not particularly well suited either, owing to the tridiagonal (recursive) algorithm, (cf. sections 5.4 and 9.2).

Recently MacCormack published a new explicit-implicit scheme |10| which is based on the original, un-split explicit method |1|. A predictor-corrector, implicit operator is incorporated into the scheme to provide the capability of taking larger steps than are allowed by the explicit stability condition. Because of the predictor-corrector sequence the coefficients of the implicit difference equations are only bi-diagonal, such that the overall effort for the complete step is roughly comparable to that of a fully implicit tridiagonal scheme for serial computers. However, MacCormack discovered that it is sufficient to consider essentially only the Euler terms of the governing equations in deriving the implicit operator using some corrections for the neglected viscous terms. Thereby the computational effort is reduced considerably compared with time-accurate implicit schemes. MacCormack shows for a scalar model equation |10| that the implicit procedures represent a perturbation of higher order small of the basic explicit scheme, and hence concludes that the total scheme remains formally second-order accurate in both, space and time. The computational effort is further reduced by the fact that the implicit sequences are only applied where it is really necessary, and are otherwise skipped completely. More recently Hung has used his experience with the time split explicit scheme to modify MacCormack's explicit-implicit scheme |11|. Thus the amount of storage needed could be reduced considerably.

This paper reports on current efforts of the author to vectorize a code, based on MacCormack's explicit-implicit scheme, for use on a CRAY-1S machine. First the governing equations are given, and the numerical scheme is sketched briefly. A discussion of the measures used to vectorize follows, and the achieved computation times on the CRAY as well as compared with that on an IBM 3081K computer are given. The sample calculation used in the comparison concerns the steady, laminar, transonic flow past a hemisphere-cylinder configuration at a free stream Mach number of 0.9, at an angle of attack of 5 degrees, and for a Reynolds number of 212500 referenced with the sphere's radius |12|.

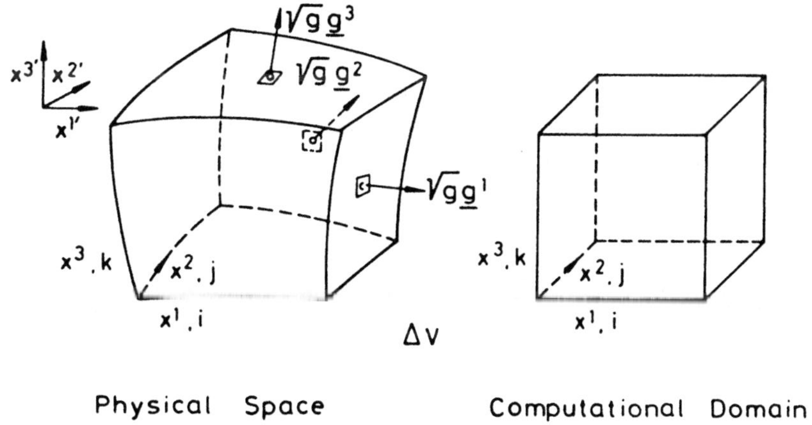

Fig. 1: Sketch of the coordinate system.

9.1.2 Governing Equations

The governing equations for Newtonian fluid flow in integral form, considering only pressure and viscous forces, for a volume V with the boundary ∂V and with the computational coordinates $\tau = t$, $x^i = x^i(x^{j'}, t)$; $i, j' = 1(1)3$, are, see Figure 1 (see also |13|):

$$(1) \quad \frac{\partial}{\partial \tau} \int_V \underline{q}\, (g)^{1/2}\, d\nu + \oint_{\partial V} (\underline{q}\underline{v} + \underline{\underline{b}}) \cdot \underline{n} dA = 0 ,$$

where all quantities are arbitrarily nondimensionalized with the corresponding free stream quantities, and where

$$\underline{q} = (\rho, \rho u, \rho v, \rho w, e)^T , \quad e = f_{pe} \frac{p}{\gamma - 1} + f_c \frac{\rho}{2} (\underline{u})^2 ,$$

$$f_{pe} \equiv 2(\gamma-1)/[2+\gamma(\gamma-1)M_\infty^2] , \quad f_c \equiv \gamma M_\infty^2 f_{pe} , f_{pu} = (\gamma M_\infty^2)^{-1} ,$$

$$\underline{v} = \underline{u} - \underline{u}_{mesh} ,$$

$$\underline{\underline{b}} = (\underline{b}_\rho, \underline{\underline{b}}_m, \underline{b}_e)^T ,$$

$$\underline{b}_\rho = 0 , \quad \underline{\underline{b}}_m = f_{pu}\, p\, \underline{\underline{I}} + Re^{-1} \underline{\underline{\tau}} ,$$

$$\underline{\underline{\tau}} = -\lambda \operatorname{div} \underline{u} \underline{\underline{I}} - \mu [(\operatorname{grad} \underline{u}) + (\operatorname{grad} \underline{u})^T] ,$$

$$\underline{b}_e = -\gamma (Re\, Pr)^{-1} \mu \operatorname{grad} T + f_{pe}\, p \underline{u} + f_c Re^{-1} \underline{\underline{\tau}} \cdot \underline{u} .$$

Here $(g)^{1/2}$ is the volume of a computational cell $d\mathcal{V}$ or the inverse of the Jacobian of the transformation, for its efficient computation see |14|. In finite-volume formulations, which are used here, the transformations are not needed explicitly, and serve solely to conveniently generate meshes and to provide an ordering of the solution sweeps. The usual notation for the flow quantities is used. Cartesian velocity vectors have been chosen in the solution vector to reduce the computational effort. The partial differential equation corresponding to equation (1) is

$$\frac{\partial}{\partial \tau} \underline{q}(g)^{1/2} + \frac{\partial}{\partial x^{\ell}} [(\underline{q}\underline{v} + \underline{b})(g)^{1/2} \underline{g}^{\ell}] = 0$$

(2)

$$\hat{\underline{q}},_{\tau} + {}^{\ell}\hat{\underline{F}},_{\ell} = 0$$

where $g^{1/2} g^{\ell}$ denotes the contravariant surface normals of surfaces x^{ℓ} =constant. Differentiated with respect to time, equation (2), with the viscosity term neglected, provides the basis for the implicit algorithm |10|

(3) $$\tilde{\hat{\underline{q}}},_{\tau} + ({}^{\ell}\hat{\underline{\underline{A}}}\tilde{\hat{\underline{q}}}),_{\ell} = 0 , \qquad \tilde{\hat{\underline{q}}} \equiv \hat{\underline{q}},_{\tau} , \qquad {}^{\ell}\hat{\underline{\underline{A}}} \equiv \frac{\partial\, {}^{\ell}\hat{\underline{F}}}{\partial \hat{\underline{q}}} ,$$

to be discussed in the next chapter. All equations have to be supplemented with appropriate initial and boundary conditions, and with the constitutive, thermodynamic relations, as well as eventually with a turbulence model.

9.1.3 The Explicit-Implicit Algorithm

Let $\Delta\tau$ be the time step, n the time level, $\underline{U}$ the numerical solution vector, ${}^{\ell}\hat{\underline{F}}$ the flux across the cell surface x^{ℓ}= constant and L_{ℓ} the one-dimensional factors of the implicit operator with respect to the direction x^{ℓ}. The predictor-corrector sequences then take the following form:

PREDICTOR:

$$\Delta\underline{U}^{n} = -\Delta\tau(\Delta_{+}\, {}^{\ell}\hat{\underline{F}}/\Delta x^{\ell}) , \qquad \ell = 1(1)3,$$

(4)

$$(L_1 L_2 L_3)_{+}\, CFL_2\, \delta\bar{\underline{U}} = CFL_2\, \Delta\underline{U}^{n} ,$$

$$CFL_1 \equiv \min[1.0, 0.5CFL/CFL_{i,j,k}] , CFL_2 \equiv 1\text{-}CFL_1 ,$$

$$\bar{\underline{U}} = \underline{U}^{n} + CFL_2\, \delta\bar{\underline{U}} + CFL_1\, \Delta\underline{U}^{n} .$$

CORRECTOR:

$$\Delta\bar{\underline{U}} = -\Delta\tau(\Delta_{-}\, {}^{\ell}\bar{\hat{\underline{F}}}/\Delta x^{\ell}) , \qquad \ell = 1(1)3,$$

(5)

$$(L_1 L_2 L_3)_{-}\, \overline{CFL_2}\, \delta\underline{U}^{n+1} = \overline{CFL_2}\, \Delta\bar{\underline{U}} ,$$

$$\underline{U}^{n+1} = \frac{1}{2} [\underline{U}^{n} + \bar{\underline{U}} + \overline{CFL_2}\, \delta\underline{U}^{n+1} + \overline{CFL_1}\, \Delta\bar{\underline{U}}] ,$$

where the subscripts indicating the spatial location have been omitted for clarity. The plus and minus subscripts indicate forward and backward two-point differences in positive or negative direction x^{ℓ}. The bars in the corrector sweeps indicate that the quantities are determined with predictor values. Note that the sweeping directions for the x^1- and x^2-sweeps are reversed every other time step.

The above scheme differs from the original one in |10| due to the weighting of explicit and implicit solutions, providing a smooth blending from one to the other. This proved to be quite useful for the computation of 2-D transonic flow |13|. The value of CFL in the weighting terms is 0.9 and $CFL_{i,j,k}$ is the local three-dimensional Courant-Friedrichs-Lewy number.

In order to heuristically obtain an idea of the explicit sweeps, consider the integral equation (1) for a single cell i with volume ΔV_i with respect to one direction x^i only, see Fig. 2. Using the abbreviations introduced in equation (2) and the notation ${}^{\ell}\underline{\underline{G}}\,(\underline{n}\,dA) = {}^{\ell}\hat{\underline{F}}$, equation (1) becomes:

$$(6) \qquad \Delta\hat{q}_{i+1/2} = -\Delta\tau\left[{}^{\prime}\underline{\underline{G}}_{i} \cdot (\underline{n}dA)_{i} + {}^{\prime}\underline{\underline{G}}_{i+1} \cdot (\underline{n}dA)_{i+1}\right] ,$$

and the predictor-corrector sweeps are as follows:

$$(7) \qquad \begin{aligned} P: \quad & \Delta\bar{\hat{q}}_{i+1/2} = -\Delta\tau\left[{}^{\prime}\underline{\underline{G}}_{i+3/2} \cdot (\underline{n}dA)_{i+1} + {}^{\prime}\underline{\underline{G}}_{i+1/2} \; (\underline{n}dA)_{i}\right] , \\ C: \quad & \Delta\hat{q}_{i+1/2} = -\Delta\tau\left[{}^{\prime}\underline{\underline{G}}_{i+1/2} \cdot (\underline{n}dA)_{i+1} + {}^{\prime}\underline{\underline{G}}_{i-1/2} \; (\underline{n}dA)_{i}\right] . \end{aligned}$$

Using the non-rigorous argument of simply adding the two contributions with the further assumption of constant matrices, one immediately recognizes a second-order accurate approximation of the fluxes at the control surfaces in equation (6). For the implicit sweeps, one typically employs the following approximation to equation (3) with respect to one direction, e.g. x^1, see Fig. 2:

$$
(8)\quad
\begin{aligned}
P:&\quad \left(\underline{\underline{I}} + \frac{\Delta\tau}{\Delta x^1_+}\,{}^1\underline{\underline{C}}_{i+1/2}\right)\delta\bar{\underline{U}}^1_{i+1/2} = CFL_2\,\Delta\underline{U}^n + \frac{\Delta\tau}{\Delta x^1_+}\,{}^1\underline{\underline{C}}_{i+3/2}\,\delta\bar{\underline{U}}^1_{i+3/2}\,,\\
C:&\quad \left(\underline{\underline{I}} + \frac{\Delta\tau}{\Delta x^1_-}\,{}^1\bar{\underline{\underline{C}}}_{i+1/2}\right)\delta\underline{U}^1_{i+1/2} = \overline{CFL_2\,\Delta\bar{\underline{U}}^n} + \frac{\Delta\tau}{\Delta x^1_-}\,{}^1\underline{\underline{C}}_{i-1/2}\,\delta\underline{U}^1_{i-1/2}\,.
\end{aligned}
$$

where the bi-diagonal structure is obvious, and where the matrices ${}^1\underline{\underline{C}}$ are obtained from the Jacobians of the Euler equations with respect to the direction x^1, however with modified, positive definite eigenvalues |10|. The control surfaces considered in the given predictor and corrector sweeps are taken at i and (i+1), and (i+1) and i, respectively, see Fig. 2. The explicit stability condition is checked for each cell i, before the matrices ${}^1\underline{\underline{C}}$ are determined, and an inversion is only performed where necessary |10|.

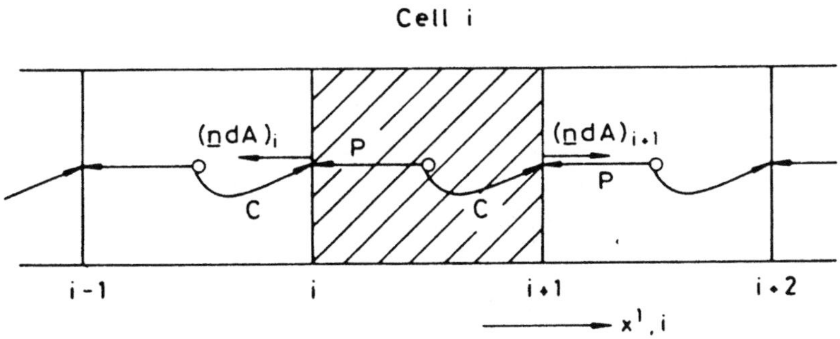

Fig. 2: Sketch of the fluxes involved in the balance for one cell.
P: Predictor, C: Corrector.

9.1.4 Vectorization

Several authors have successfully vectorized MacCormack's explicit scheme for use on a CRAY-1S computer |15,16|. For the computation of a three-dimensional, turbulent, transonic diffuser flow, Shang et al. |15| achieved a reduction of computation time (expressed in terms of Rate of Data Processing: RDP = CPU time/ (total number of grid points times total number of iterations)) of roughly 128 compared to the time spent on a CYBER 74 computer. For later comparison it is noted that this achievement corresponds theoretically to a factor of about 37 for the IBM 3081K, assuming that the factor between IBM and CDC 7600 is about 2/3 *), and using the timing for the CDC 7600 as given in reference |15|.

For the calculation of inviscid and (based on the thin-layer Navier-Stokes approximation, i.e. retaining only the boundary-layer terms in the diffusion terms) laminar flows past bicircular arc airfoils Chima and Johnson |16| report a decrease in time by a factor of about 13 and 20 (viscous) with respect to the time on an IBM 370/3033 computer (corresponding to about 7.5 and 14, respectively, for the IBM 3081K assuming a ratio of about 1.4 between IBM 3081K and 3033U *)). Note that any such factors given have to be regarded with particular caution, since they depend strongly upon the adaptation of the original code to the scalar machine used for comparison. Also the factors between the present IBM 3081K and the other computers can only be regarded as tentative since those numbers are highly code and case dependent.

Before the current achievements are to be discussed, it is useful to know the structure of the time-consuming parts of the original serial code which has proved quite favourable to vectorization. Both, explicit and implicit, predictor and corrector sweeps, equations (7) and (8), are performed, each in a separate subroutine: EXPL-n and IMPL-n for each direction x^n, n=1(1)3, considered, see Fig. 1. Each of the six subroutines consists essentially of three nested DO-loops, namely for the x^1-, x^2- and x^3-direction, with the n-loop as innermost loop. Within this inner loop the flux across one surface of cell n is determined for each value of n, see Figure 2. In the explicit routines the loop has always the range from n=1 to n=(nmax-1), and uses the surface normals at x^{n+1}=constant; the appropriate predictor or corrector fluxes, see Figure 2 for n=i, are obtained by adding or not an increment of one to n to fetch the

---

*) This ratio is based on discussions with W. Gentzsch and D. Mueller-Wichards, DFVLR, Goettingen. The ratio 2/3 between the 3081K and CDC 7600 has been derived from benchmark runs with fundamental sample codes as conducted by a DFVLR group |17|. The ratio of 1.4 for one processor of the 3081K versus the one processor machine 3033U is obtained from scientific benchmark runs made by IBM.

corresponding dependent variables. The flux-differencing is done with the help of a small two-dimensional array FLX(5,2) for the fluxes on both sides of the cell n, and of an alternating algorithm to assign the values of 1 or 2 to m in Flx(5,m). The numerical diffusion terms are added within the DO-loop if wanted. In the implicit routines the DO-loop variable follows the chosen sweeping direction and varies from 2 to (nmax-1) or from (nmax-1) to 2, respectively. This is achieved by using a switching algorithm for the DO-loop variable n. For each value of n the corresponding dependent variables are fetched, and the appropriate surface is taken by adding an increment of one to n or not. Note that for each value of n an IF-statement is necessary to determine whether or not the implicit procedure must be followed. The fluxes on the right-hand side of equations (8) can be considered as local flux boundary condition and are determined at the end of the nth step by simply solving the predictor and corrector difference equations for, e.g., $\left| \frac{\Delta \tau}{\Delta x_+^i} \, {}^1\underline{\underline{C}}_{i+\frac{1}{2}} \, \delta \underline{U}^1_{i+\frac{1}{2}} \right| / \Delta x_+^1$, since that term multiplied by $\Delta x_+^i / \Delta x_+^{i-1}$ represents the new boundary condition in the next step of the predictor sweep. This is a form of recursion. IF numerical diffusion terms have been added in the explicit portion, such terms are used in the implicit routines, too. Hence, large DO-loops containing a moderate number of IF-statements need to be vectorized, where according to the flow trace analysis most of the computation time is spent. In the implicit routines the recursive feature has to be circumvented.

The current efforts to completely vectorize the explicit portions of the explicit-implicit scheme use the auto-vectorization feature of the CRAY compiler. Hence it was necessary to decompose long DO-loops into smaller vectorizable loops, to remove IF- and CALL-statements from DO-loops, to unroll short inner DO-loops to enable larger vector loops, and to use the direction with the largest number of points as innermost DO-loop variable. It is emphasized, however, that to produce long vector strings is not as important for a CRAY-1S as it is, for example, for a CYBER 205 computer. The introduction of dummy arrays is necessary to store intermediate scalars into vector strings which allows to exploit the vectorizing capabilities of the CRAY-1S. On the other hand, small, locally used arrays, such as FLX(5,2), are redefined into longer vector strings, for example into FLX(31,5), so that the flux-differencing in equation (7) can be performed efficiently in a vector loop. Since initially, the innermost DO-loops were not vectorized by the compiler, they are decomposed typically into the following segments: determine convective surface velocities for all n; determine the fluxes of the inviscid portion of the governing equations; determine components of g^{ℓ} for one cell, l=1(1)3; determine $\partial u^{i'} / \partial x^{\ell}$, i', ℓ = 1(1)3; determine $\partial u^{i'} / \partial x^{k'}$, i', k' =1(1)3, and the temperature; determine the heat flux in three directions; and, finally, compute the viscous fluxes and add them to the Euler fluxes simultaneously. Add numerical diffusion if necessary, and then perform the flux-differencing. This is a straight forward procedure and does not require sophisticated vector directives. The portability of the code is hampered only due to a conditional merge directive to vectorize an IF-statement involving vectors, this being necessary for stability reasons for MacCormack's method if an accelerated flow has to be simulated. Then an average value of both convective surface velocities of a cell has to be used instead of one cell value |1|. Another difficulty may arise on other computers by the used BUFFER-IN and -OUT routines as offered by the CRAY. The latter can quickly be exchanged for unformatted READ and WRITE of a large array using only the name without any DO-loop at all.

Using the above strategy the recursive bi-diagonal implicit sweeps too have been almost completely vectorized. The only exceptions are the three IF-statement loops necessary to decide whether and where the scheme is implicit at all (see above). While on serial computers it is quite reasonable to detect single locations, if multiple implicit spots occur in one direction, here strings of locations are formed by searching for the first and last "implicit point" in each direction considered. Thereby a more efficient vector loop is created and start-up times are reduced. Since the number of implicitly treated cells is printed for each direction separately, it could be checked that those numbers did not increase dramatically with respect to the runs with the original unvectorized version for the step sizes chosen. Vector loops are introduced in one of those directions which do not coincide with the recursive direction. For different meshes one has then to take care that a direction is used which produces longest vector strings and the smallest number of DO-loops to be initiated in order to reduce the start up time for the initialization of the vector loop. One could also use complete vectorization for two coordinate directions, corresponding to one entire plane of data in the computational domain. This, however, would require more space for dummy arrays, and reduce the extremely valuable space for in-core storage of data for three dimensions. An alternative would be to use external storage devices and to keep only three surfaces of data in core at one time, but this would require efficient input/output devices (see also |15|). The segmentation is not explained in detail here as it is again straight forward.

Owing to the CFL-weighting, see equations (3) and (4), additional three-dimensional arrays for each of the five dependent variables would have been necessary. Instead, writing on and reading from temporary disk files has been used in order to maintain minimal in-core storage. BUFFER-IN and BUFFER-OUT statements proved to reduce the time, required initially for shuffling the data by means of unformatted READ or WRITE of one single huge array by more than sixty per cent. This is one of the finer improvements, however, having only a small effect on the overall time.

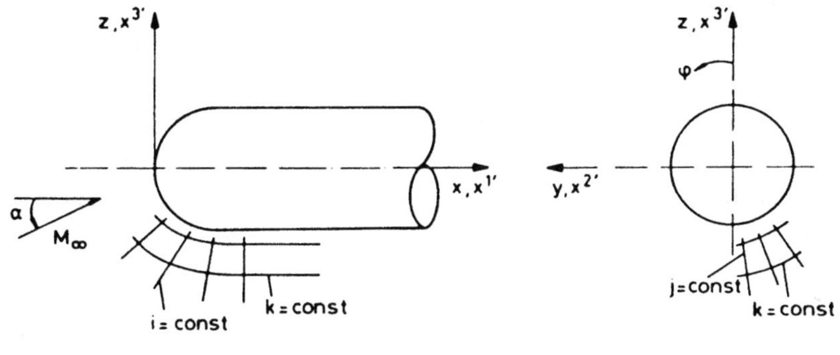

Fig. 3: Sketch of the hemisphere-cylinder configuration.

In the following the increase in efficiency is mostly expressed in terms of the above mentioned rate of data processing RDP |15|, however, using the total number of cells instead of mesh points. The present code, of course, has no claim to having reached a state of optimum vectorization, since for that a deeper insight into the architecture is required, and more time would have to be invested without substantially gaining speed, for sure. For the original code the ratio of computer times on the IBM 3081K and on the CRAY-1S using scalar performance only, is roughly 7.8, which increases to about 10 when auto-vectorization is permitted. In comparison with the hand-vectorized version (see above), a ratio of about 31.4 is achieved for the computation of laminar flow. Note that this ratio is for a slightly implicit version for the 31x20x31 mesh, where in the axial (31) direction the integration was completely explicit, and in the circumferential (20) direction the scheme was only slightly implicit (see Figure 3 for a sketch of the configuration). The ratio of 31.4 for the case of a laminar flow is about 15 per cent below the theoretical value of 37 obtained in reference 15 for the case of a turbulent flow, but larger than that one obtained in reference 16 for laminar flow, both for explicit calculations. It is not easy to give a satisfactory explanation both for these differences. Looking at the values of RDP in Table 1 for the explicit version (obtained by removing the CFL-weighting routine and the CALL-IMPL-n statements) the timing is the same as in reference 15, although without turbulence modeling the explicit version is expected to be considerably faster. This points to the implicit portion as the reason for the unfavourable comparison. Table 1 indicates that the increase of computational work due to the viscous effects of the thin-layer approximation is less than 20 per cent of the work for the Euler equations. If a scalar turbulence model (Baldwin-Lomax) is employed this results in an additional RDP of 0.0000122 for the full evaluation of the vorticity vector and in an additional RDP of 0.0000102 if the thin-layer approximation is introduced in the formula for the vorticity vector. These numbers hold when the corresponding subroutine TURBV to calculate the eddy viscosity is called twice per time step, after both, predictor and corrector sweeps. It is emphasized that, for turbulent flows, the above numbers give only a hint to the time requirements, because in the turbulent flows the solution may lead to a larger number of implicitly treated cells, and may thus change somewhat the times displayed in Table 1. Note that the first step sizes away from the body in terms of radii, are fairly small, namely 0.00005 for the viscous (laminar) and 0.005 for the inviscid case. Table 1 shows to what extent the time requirements increase if the full Navier-Stokes solution is used instead of the thin-layer approximation. The increase is largest for the explicit calculations while the effect is less dramatic in the case of implicit integration. Note that the time requirements (per cell and time step) for the implicit solution remain nearly the same if the grid size is increased by roughly thirty per cent.

Table 1: Computation times for the vectorized Euler and laminar Navier-Stokes solutions (per mesh cell and iteration, in seconds).

	Mesh	31 x 20 x 31 (19220) *			42 x 20 x 31 (26040)	
	Method / RDP	Explicit	Implicit less	Implicit more	Implicit	
Navier - Stokes	full	5.9	8.5	9.3	8.5	10^{-5} Seconds
	CFL_i		0.23 (0)	0.56 (5168)	0.24 (0)	
	CFL_j		1.24 (1207)	3.03 (2139)	1.28 (1139)	
	CFL_k		164 (8349)	390 (9543)	167 (11520)	
	thin-layer	2.5	5.7	6	5.5	10^{-5} Seconds
	CFL_i		0.23 (0)	0.56 (5168)	0.24 (0)	
	CFL_j		1.24 (1270)	3.03 (2139)	1.28 (1139)	
	CFL_k		164 (8349)	390 (9543)	167 (11520)	
Euler		2.1	4.9	5.1	5	10^{-5} Seconds
	CFL_i		0.22 (0)	0.40 (1366)	0.47 (3735)	
	CFL_j		1.22 (1903)	2.22 (1377)	2.54 (1288)	
	CFL_k		1.56 (3842)	2.85 (5418)	3.29 (7600)	

*) (...): number of cells, in the first row this includes boundary cells

CFL_n: maximum CFL-number in direction x^n

RDP : CPU-time/(cells * iterations)

9.1.5 Concluding Remarks

The development of MacCormack's predictor-corrector methods from 1969 to present has been sketched, and his latest method, an explicit-implicit scheme, has been described for application in three dimensions and general coordinates. The scheme has been shown to be amenable to efficient vectorization on the CRAY-1S using the compiler's auto-vectorization feature, i.e. by breaking up the code in vectorizable DO-loops. For a slightly implicit case, a rate of data processing (defined as total CPU-time per (total number of cells times total number of iterations)) of about $9.1*10^{-5}$ seconds is achieved. This enables the efficient tackling of those problems whose solutions require a finer mesh, provided efficient input/output devices in conjunction with a fast external memory are available..

References

1. MacCormack R.W.: The effect of viscosity in hypervelocity impact cratering. AIAA Paper 69-354, 1969.
2. MacCormack R.W.: The numerical solution of the interaction of a shock wave with a laminar boundary layer. Proceedings, 2nd Int. Conf.Numerical Methods in Fluid Dynamics, Lecture Notes in Physics, Vol. 8, Springer Verlag, Berlin 1971, 151-163.
3. MacCormack R.W., Baldwin B.S.: A numerical method for solving the Navier-Stokes equation with application to shock-boundary-layer interactions. AIAA Paper 75-1, 1975.
4. Rizzi A.W.,Inouye M.: A time-split finite-volume technique for the three-dimensional blunt-body flow. AIAA Paper 73-133, 1973.
5. Kutler P., Reinhardt W.A., Warming R.F.: Numerical computation of multishocked,three-dimensional supersonic flow fields with real gas effects. AIAA Paper 72-192, 1972.
6. Deiwert G.S.:Numerical simulation of high Reynolds number transonic flows. AIAA Journal, Vol. 13, pp. 1354-1359, 1975.
7. MacCormack R.W.: An efficient numerical method for solving the time dependent compressible Navier-Stokes equations at high Reynolds number. NASA TM X-73,129, July 1976.
8. Hung C.M., MacCormack R.W.: Numerical solution of three-dimensional shock wave and turbulent boundary-layer interaction. AIAA Paper 78-161, 1978.
9. MacCormack R.W., Lomax H.: Numerical solution of compressible viscous flows. Annual Review of Fluid Mechnics, Vol. 11, 1979, 289-316.
10. MacCormack R.W.: A numerical method for solving the equations of compressible viscous flow. AIAA Paper 81-110, 1981 (see also AIAA Journal, Vol 20, No 9, Sept. 1982, pp 1275-1281).
11. Hung C.M.,Kordulla W.K.: A time-split finite-volume algorithm for three-dimensional flow-field simulation. AIAA Paper 83-1957, 1983.
12. Kordulla W.: The computation of three-dimensional transonic flows

with an explicit-implicit method. Proceedings, 5th GAMM-Conf. Numerical Methods in Fluid Dynamics, Rome 1983. Notes on Numerical Fluid Dynamic, Vol.7 (Pandolfi M., Piva R.,eds.), Vieweg Verlag, Braunschweig 1984.

13. Kordulla W., MacCormack R.W.: Transonic-flow computations using an explicit-implicit method. Proceedings, 8th Int. Conf. Numerical Methods in Fluid Dynamics, Lecture Notes in Physics, Vol. 170, Springer Verlag, 1982, 420-426.
14. Kordulla W., Vinokur M.: Efficient computation of volume in flow predictions. AIAA Journal, Vol. 21, 1983, 917-918.
15. Shang J.S.,Buning P.G.,Hankey W.L.,Wirth M.C.: The performance of a vectorized 3-D Navier-Stokes code on the Cray-1 computer. AIAA Paper 79-1448,1979 (see also AIAA Journal, vol 18, Sept. 1980, 1073-1078).
16. Chima R.V., Johnson G.M.: Efficient solution of the Euler and Navier-Stokes equations with a vectorized multiple-grid algorithm. AIAA Paper 83-1893, 1983.
17. Mueller-Wichards D. and Gentzsch W.: Performance comparisons among several parallel and vector computers on a set of fluid flow problems. DFVLR-IB 262-82 R 01, 1982.

9.2 Vectorization of the Implicit Beam and Warming Scheme

Bernhard Mueller, DFVLR Goettingen *

9.2.1 Introduction

The investigation of vortical flows using the Euler equations and the study of viscous-inviscid interactions using the Navier-Stokes equations have gained increased interest among aerodynamicists. In recent years, implicit methods have been used in increasing number in the numerical simulation of compressible flows. In this article, it will be shown how the implicit Beam and Warming scheme |1| can be implemented on a vector computer. This finite-difference method is easily vectorizable except for the solution of the resulting block-tridiagonal linear systems. By applying the Richtmyer algorithm simultaneously instead of separately, the block-tridiagonal system solver can also be vectorized, although at the cost of increased storage requirements. The advantages of vectorization are demonstrated for the computed example of axisymmetric laminar supersonic flow over a hemisphere-cylinder.

In the present analysis, the thin-layer approximation of the Navier-Stokes equations is employed (cf. chapter 9.2.2) to examine high Reynolds number flows. In chapter 9.2.3, the implicit factored finite-difference scheme of Beam and Warming is formulated. The algebraic grid generation, the initialization and the boundary treatment using shock fitting, similar to the approach of Kutler et al. |2|, are described in chapter 9.2.4. Results are compared with other numerical approaches (cf. chapter 9.2.5) for axisymmetric laminar supersonic flow over a hemisphere-cylinder. In chapter 9.2.6, the vectorization of the implicit algorithm is illustrated for the CRAY-1S. As the solution of the block-tridiagonal linear systems constitutes the most time consuming part of the computer program, a vectorizable method, namely the simultaneous treatment of the Richtmyer algorithm, is stated in chapter 9.2.7. In chapter 9.2.8, the advantages of vectorization are discussed.

*) This research was partially supported by the Deutsche Forschungsgemeinschaft under the program "Finite Approximationen in der Stroemungsmechanik".

9.2.2 Governing Equations

The independent variables of the time-dependent compressible Navier-Stokes equations written in Cartesian coordinates are transformed |3| (cf. Fig.1 for the problem of interest) by

$$\tau = t$$

$$\xi = \xi(t,x,y) \; ,$$

$$\eta = \eta(t,x,y) \; .$$

Employing the thin-layer approximation |4| in the resulting equations implies that the body surface conforms to a line of constant η, and that among the viscous terms only gradients in the near normal η-direction need to be resolved.

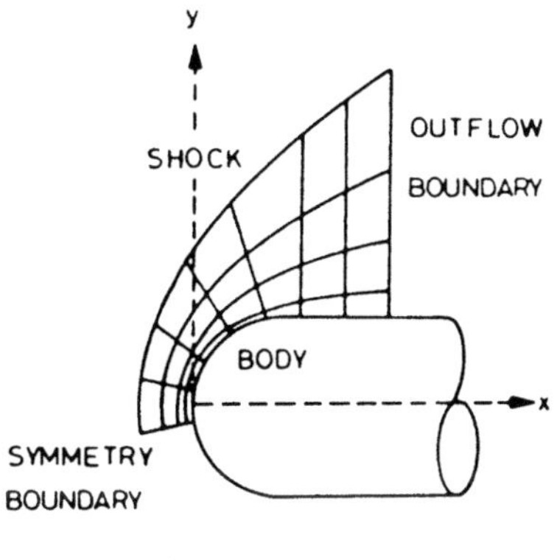

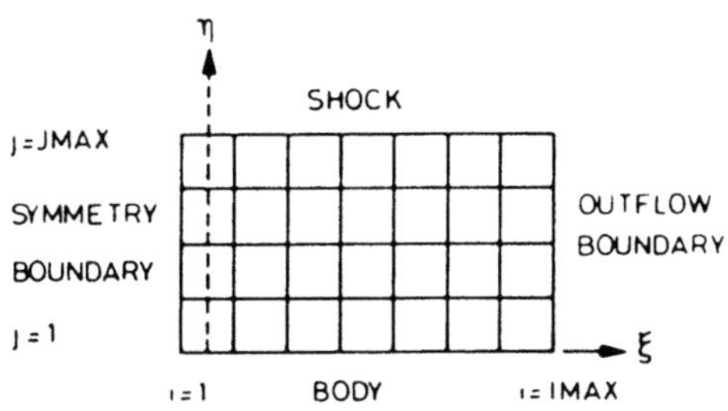

Fig. 1: Transformation of physical plane to computational plane.

For axisymmetric flow of perfect gas with constant specific heats and without external forces, the time-dependent compressible thin-layer Navier-Stokes equations may be written in the dimensionless conservation - law form |2| as:

$$(1) \qquad \frac{\partial q}{\partial \tau} + \frac{\partial E}{\partial \xi} + \frac{\partial F}{\partial \eta} + H = \frac{1}{Re}\left[\frac{\partial S}{\partial \eta} + T\right]$$

where

$$q = \frac{1}{J}\begin{pmatrix}\rho\\ \rho u\\ \rho v\\ e\end{pmatrix}, \quad E = \frac{1}{J}\begin{pmatrix}\rho U\\ \rho uU+\xi_x p\\ \rho vU+\xi_y p\\ (e+p)U-\xi_t p\end{pmatrix}, \quad F = \frac{1}{J}\begin{pmatrix}\rho V\\ \rho uV+\eta_x p\\ \rho vV+\eta_y p\\ (e+p)V-\eta_t p\end{pmatrix}, \quad H = \frac{v}{Jy}\begin{pmatrix}\rho\\ \rho u\\ \rho v\\ e+p\end{pmatrix},$$

$$S = \frac{1}{J}\begin{pmatrix}0\\ \eta_x\tau_{xx} + \eta_y\tau_{xy}\\ \eta_x\tau_{xy} + \eta_y\tau_{yy}\\ \eta_x(u\tau_{xx} + v\tau_{xy} + \frac{\gamma\kappa}{Pr}\eta_x\frac{\partial\varepsilon}{\partial\eta}) + \eta_y(u\tau_{xy} + v\tau_{yy} + \frac{\gamma\kappa}{Pr}\eta_y\frac{\partial\varepsilon}{\partial\eta})\end{pmatrix},$$

$$T = \frac{1}{Jy}\begin{pmatrix}0\\ \tau_{xy}\\ \tau_{yy} - \tau_{\theta\theta}\\ u\tau_{xy} + v\tau_{yy} + \frac{\gamma\kappa}{Pr}\eta_y\frac{\partial\varepsilon}{\partial\eta}\end{pmatrix},$$

with

$$U = \xi_t + \xi_x u + \xi_y v ,$$

$$V = \eta_t + \eta_x u + \eta_y v ,$$

$$\tau_{xx} = (\lambda+2\mu)\,\eta_x\frac{\partial u}{\partial\eta} + \lambda\eta_y\frac{\partial v}{\partial\eta} + \frac{\lambda v}{y} ,$$

$$\tau_{xy} = \mu(\eta_y\frac{\partial u}{\partial\eta} + \eta_x\frac{\partial v}{\partial\eta}) ,$$

$$\tau_{yy} = (\lambda+2\mu)\,\eta_y\frac{\partial v}{\partial\eta} + \lambda\eta_x\frac{\partial u}{\partial\eta} + \frac{\lambda v}{y} ,$$

$$\tau_{\theta\theta} = (\lambda+2\mu)\,\frac{v}{y} + \lambda\,(\eta_x\frac{\partial u}{\partial\eta} + \eta_y\frac{\partial v}{\partial\eta}) .$$

The Jacobian J of the transformation is given by

$$(2) \qquad J = \frac{1}{\frac{\partial x}{\partial\xi}\frac{\partial y}{\partial\eta} - \frac{\partial x}{\partial\eta}\frac{\partial y}{\partial\xi}}$$

and the metrics are formed in terms of the derivatives of the Cartesian coordinates as follows:

$$\xi_x = J\frac{\partial y}{\partial \eta}\ , \qquad \eta_x = -J\frac{\partial y}{\partial \xi}$$

(3) $$\xi_y = -J\frac{\partial x}{\partial \eta}\ , \qquad \eta_y = J\frac{\partial x}{\partial \xi}$$

$$\xi_t = -\frac{\partial x}{\partial \tau}\xi_x - \frac{\partial y}{\partial \tau}\xi_y\ , \qquad \eta_t = -\frac{\partial x}{\partial \tau}\eta_x - \frac{\partial y}{\partial \tau}\eta_y$$

where the convention $\xi_x = \frac{\partial \xi}{\partial x}$, etc. is used.

The density ρ is made non-dimensional by its freestream value ρ_∞, the x and y velocity components u and v by u_∞, the viscosity coefficients λ and μ by μ_∞, the coefficient of thermal conductivity κ by κ_∞, the temperature T by T_∞, the total energy per unit volume e by $\rho_\infty u_\infty^2$, the x and y coordinates by a characteristic length L, and the time t by $\frac{L}{u_\infty}$. The Reynolds number Re, the Prandtl number Pr, and the Mach number M_∞ are defined in terms of these reference values. γ represents the ratio of specific heats. Normalized by $\rho_\infty u_\infty^2$, the pressure p is related to the specific internal energy ε with reference quantity u_∞^2 by the equation of state for perfect gas

(4) $$p = (\gamma - 1)\rho\varepsilon$$

where

$$\varepsilon = \frac{e}{\rho} - \frac{u^2+v^2}{2}\ .$$

Stokes's hypothesis is imployed to yield

(5) $$\lambda = -\frac{2}{3}\mu\ .$$

As the Prandtl number is assumed to be constant, the following relation holds for the dimensionless values

(6) $$\kappa = \mu\ .$$

The dependence of the viscosity coefficient μ on the temperature T is obtained from the Sutherland law

(7) $$\mu = T^{3/2}\frac{1+S_1}{T+S_1}\ , \quad \text{where } S_1 = \frac{110\,^\circ K}{T_\infty} \text{ for air.}$$

9.2.3 Algorithm

The implicit factored finite-difference scheme of Beam and Warming |1| is employed to solve the axisymmetric thin-layer Navier-Stokes equations. The time derivative is approximated by the first-order Euler implicit, i.e. $\vartheta=1$ and $\psi=0$, or the second-order three-point backward formula, i.e. $\vartheta=1$ and $\psi=0.5$ (cf. (8.a) and (8.b) below). The flux vectors are linearized by Taylor series expansions, and second-order central differences are used to evaluate the metric terms (2) and (3) and the spatial derivatives of the flow variables at interior grid points. The resulting finite-difference equation is approximately factored and implemented by a sequence of two one-dimensional matrix inversions referred to as the ξ- and the η- sweep:

ξ-sweep:

$$\left[I + \frac{\vartheta\Delta\tau}{1+\psi}\,\delta_\xi A^n - \varepsilon_I\left(J^{-1}\nabla_\xi\Delta_\xi J\right)^n\right]\Delta q^{*n}$$

(8.a)

$$= -\frac{\Delta\tau}{1+\psi}\left[\delta_\xi E^n + \delta_\eta F^n + H^n - \frac{1}{Re}\left(\delta_\eta S^n + T^n\right)\right] + \frac{\psi}{1+\psi}\Delta q^{n-1}$$

$$-\varepsilon_E\left(J^n\right)^{-1}\left[\left(\nabla_\xi\Delta_\xi\right)^2 + \left(\nabla_\eta\Delta_\eta\right)^2\right](Jq)^n \quad ,$$

η-sweep:

$$\left[I + \frac{\vartheta\Delta\tau}{1+\psi}\left(\delta_\eta\left(B^n - \frac{1}{Re}M^n\right) + \left(C^n - \frac{1}{Re}N^n\right)\right)\right.$$

(8.b)

$$\left. - \varepsilon_I\left(J^{-1}\nabla_\eta\Delta_\eta J\right)^n\right]\Delta q^n = \Delta q^{*n} \quad ,$$

(8.c)

$$q^{n+1} = q^n + \Delta q^n \quad ,$$

where A, B, C, M, and N are the Jacobian matrices of E, F, H, S, and T, resp. (cf. |2|, |23|), and I is the identity matrix. $\delta\xi$ and $\delta\eta$ are second-order central difference operators, Δ and ∇ represent the conventional forward and backward difference operators. Note that the operators enclosed in brackets are to be applied to Δq^{*n}, $(Jq)^n$, and Δq^n, resp., except for $\delta_\eta u$ and $\delta_\eta v$ in N. Whereas the inviscid flux terms, i.e. E, F, and H, are linearized to second-order temporal accuracy, the viscous ones, i.e. M and N, are linearized only to first-order |5|, because the dependence of $\frac{\mu}{J}$ on q may be neglected, if the transport coefficients are changing slowly with time. Fourth-order numerical dissipation terms with $\varepsilon_E = \Delta\tau$ are added explicitly to the equations to damp short wave length oscillations. Also the addition of the implicit second-order smoothing terms with $\varepsilon_I = 2\,\varepsilon_E$ serves to control the numerical stability (cf. |6|).

Equations (8.a), (8.b), and (8.c) are applied to the interior points only. Explicit numerical dissipation terms are cancelled at the grid points adjacent to the body, the outflow boundary and the shock. The ξ-sweep consists of forming the block-tridiagonal matrices and the right hand sides of (8.a), solving the block-tridiagonal system , and storing the result on the Δq^n array for each line of constant η . Similarly, the η-sweep consists of forming the block-tridiagonal matrices of (8.b), solving the block-tridiagonal system, and computing the solution q^{n+1} from (8.c) for each line of constant ξ . In the present case the blocks are 4x4 matrices. The computation of the block-tridiagonal matrices of the ξ- and η-sweeps and the right hand side of the ξ-sweep is easily vectorizable, because these terms depend on q^n, and on q^n and Δq^{n-1}, resp., only. If each block-tridiagonal system is solved by the Richtmyer algorithm, recursions inhibit vectorization. A vectorizable alternative will be given in chapter 9.2.7.

9.2.4 Grid Generation, Initialization, Boundary Treatment

To calculate supersonic flow over a hemisphere-cylinder, the grid is generated algebraically. The lines of constant ξ, distributed at equal arclengths normal to the body surface are kept fixed, whereas the lines of constant η, clustered by the function

$$f(\zeta) = 1+\beta\,\frac{1-\left(\frac{\beta+1}{\beta-1}\right)^{1-\zeta}}{1+\left(\frac{\beta+1}{\beta-1}\right)^{1-\zeta}}\;,\zeta \in [0,1]\,,\beta>1\quad,$$

(cf. |7|) near the body are moved according to the shock position. The initial shape and position of the shock are prescribed according to |8|. On the body surface, a modified Newtonian theory is used to prescribe the initial pressure distribution (cf. |9|). For an adiabatic wall, the temperature is initialized by the stagnation point temperature. In addition the no-slip condition is imposed. Assuming a zero shock velocity, the flow variables behind the shock are obtained from the freestream conditions by employing the Rankine-Hugoniot relations. The flow variables at the interior grid points on the lines of constant ξ are equated to their values downstream of the shock.

If the body surface is not moving, the no-slip condition implies $u = v = 0$. Neglecting the viscous terms in the normal momentum equation leads to a zero normal pressure gradient at the surface. For an adiabatic wall the normal wall temperature gradient is also zero.

The first two lines of constant ξ are chosen to straddle the axis of symmetry in order to allow the application of the reflection principle at the symmetry boundary |2|.

Since the last line of constant ξ is chosen such that the normal velocity is supersonic along that outflow boundary except in the vicinity of the body surface, a simple linear extrapolation of the conserved variables $\rho, \rho u, \rho v$, and e is used. The pressure behind the shock, where the flow is assumed to be inviscid, is determined from the compatibility equation corresponding to the only positive eigenvalue $V + c\sqrt{\eta_x^2+\eta_y^2}$ of the Jacobian matrix B where c is the speed of sound |10|. Then the other flow variables and the shock velocity are evaluated from the Rankine-Hugoniot relations |11|,|2|,|10|.

The boundary conditions at the body and the shock, where the ξ- and η-derivatives are approximated by second-order differences, are treated explicitly, and those at the symmetry boundary and the outflow boundary are imposed implicitly.

The shock position at the new time level is determined by using the shock velocity to propagate the shock along the lines of constant ξ (cf. |12|).

9.2.5 Results

To verify the blunt body code, laminar flow over a hemisphere-cylinder with an adiabatic wall at $M_\infty = 2.94$, $Re = 2.2 \times 10^5$, $\gamma = 1.4$, $Pr = 0.723$, $T_\infty = 293^\circ K$ was chosen |23|. The grid consisted of 26 points in the ξ-direction and 31 points in the η-direction. For the first 100 time levels, the time step was linearly increased to its final value, which corresponded to a maximum Courant number of 156. After 400 time levels the steady-state solution was obtained, the convergence being estimated using $\frac{\Delta J q}{\Delta \tau}$ and the shock velocity.

The shock shape, the sonic line location and the surface pressure distribution are in good agreement with a Navier-Stokes solution |13| and with thin-layer Navier-Stokes results |2|,|12|,|14|,|15| (cf. Figures 2 and 3). The surface temperature differs from the Navier-Stokes result |13| at most by 3.5 per cent, based on the latter result (cf. Figure 4). The agreement with |2| is better than with |14|,|15|. With a finer mesh clustered closer to the body for the free parameter β of the clustering function closer to 1, the difference of the present result is at most 2.4 per cent compared with |13|. The example of 400 time levels on the 26 x 31 mesh was chosen to illustrate the vectorization of the Beam and Warming scheme and to compare the computing speed of different versions of the code on the CRAY-1S.

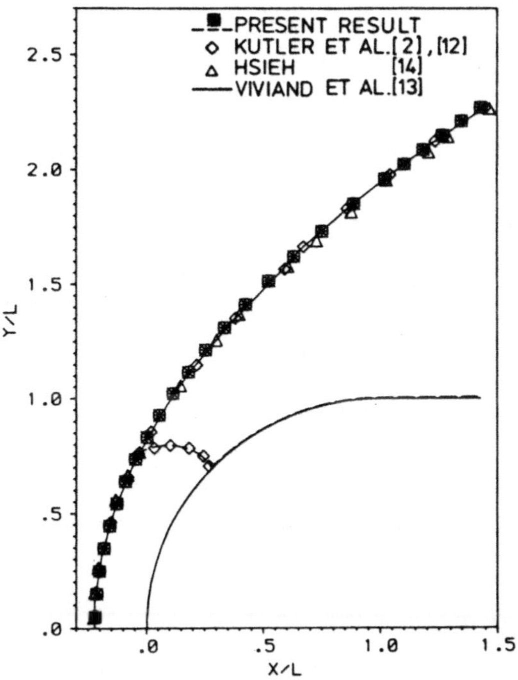

Fig. 2: Shock shape and sonic line location for hemisphere-cylinder with adiabatic wall at $M_\infty = 2.94$, $Re = 2.2*10^5$.

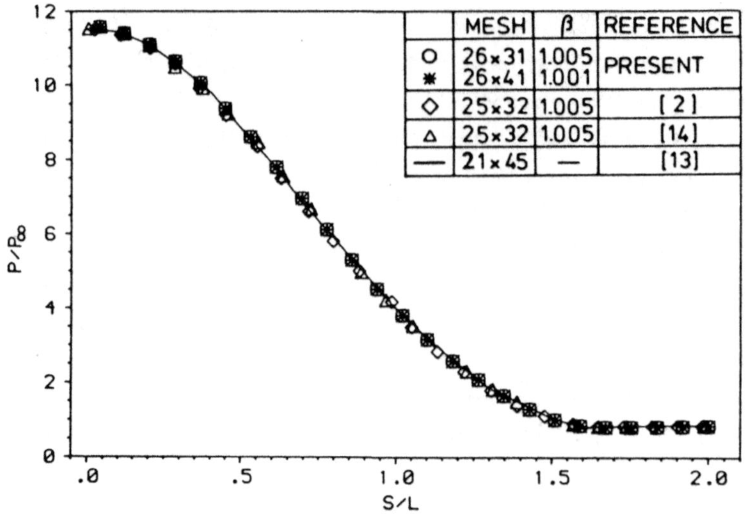

Fig. 3: Surface pressure distribution for hemisphere-cylinder with adiabatic wall at $M_\infty = 2.94$, $Re = 2.2*10^5$ (β: parameter of clustering function, cf. chapter 9.2.4, s: arclength from stagnation point).

9.2.6 Vectorization of the Beam and Warming Scheme on the CRAY-1S

The original version of the blunt body code for the solution of the axisymmetric thin-layer Navier-Stokes equations by the Beam and Warming scheme (cf. Figure 5) was written in FORTRAN IV and implemented on the IBM 3081K using the compiler option AUTODBL to change automatically from REAL*4 to REAL*8.

The implementation of the code on the CRAY-1S was performed without difficulty. 64-bit words and the compiler option ON = EFLMPQRSVZ were used. Then from the flowtrace, the most time consuming subroutines were selected and modified following the general guidelines for vectorization on the CRAY-1S |16|,|17| (cf. chapter 3.2):

(a) Remove I/O, IF, GOTO, and CALL statements and calls of FUNCTION subprograms from inner DO loops.

(b) Define and use constant integer increments properly in inner DO loops.

(c) Use as many inner DO loops as necessary and as few as possible.

(d) Unroll inner DO loops of vector lengths 2,3, and 4.

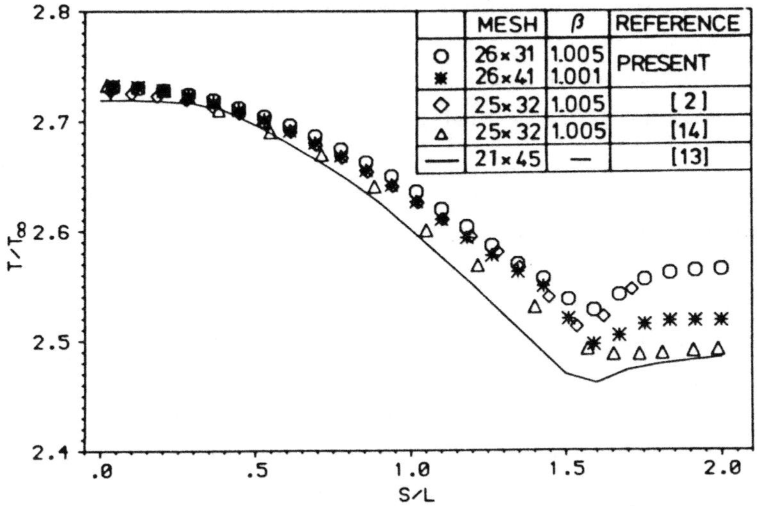

Fig. 4: Surface temperature distribution for hemisphere-cylinder with adiabatic wall at M_∞ =2.94, Re = $2.2*10^5$ (β,ς: cf. Fig. 3).

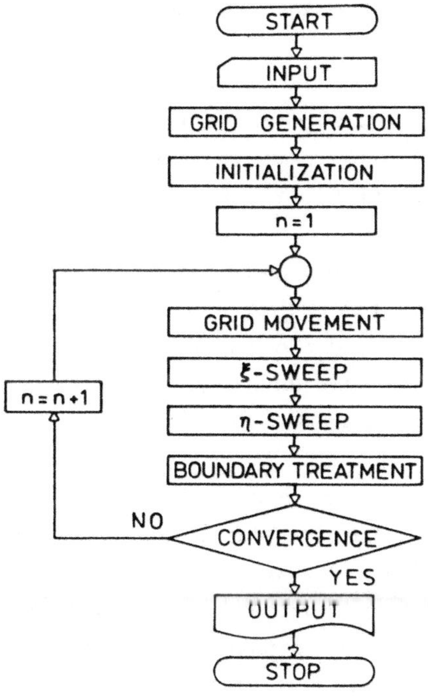

Fig. 5: Flow chart for the solution of the axisymmetric thin-layer Navier-Stokes equations by the Beam and Warming scheme.

(e) Make judicious use of the CRAY-1 intrinsic functions and utility procedures.

(f) Change the dimension to eliminate memory conflicts.

(g) Remove dependencies from inner DO loops.

For the vector-coding of the Beam and Warming scheme on the CRAY-1S, these guidelines have the following implications:

(a) None of the statements precluding vectorization was used in the subroutines forming the block-tridiagonal linear systems of the ξ- and η-sweeps.
In the subroutines solving each system by the Richtmyer algorithm, the inversion of the diagonal matrices was originally performed in two subroutines decomposing and solving the corresponding linear systems, respectively. Pulling these subroutines, which were called nearly 2.7 million times in the example, into the calling routine, increased the number of statements, but decreased the computing time of the whole block-tridiagonal inversion process by a factor of 1.36.
In the subroutine checking the convergence, the IF statements used in an inner DO loop to determine the indices where $|\frac{\Delta\rho^n}{\Delta\tau}|$, $|\frac{\Delta\rho u^n}{\Delta\tau}|$, $|\frac{\Delta\rho v^n}{\Delta\tau}|$, $|\frac{\Delta e^n}{\Delta\tau}|$ were maximal, were removed by introducing temporary arrays and employing the CRAY-1 FORTRAN intrinsic function ISMAX.

(b) In the suboutines moving the grid, a constant integer increment was used as argument in a nested loop. Switching the loops rendered the inner one vectorizable, and allowed to remove a time consuming operation statement from the inner loop. The execution time for that subroutine was improved by a factor of 17.36.

(c) The computation of the block-tridiagonal matrices was not vectorized by the compiler because of the number and complexity of the 48 elements to be calculated in a single DO loop. The remedy was to split the loop into three, and to calculate common temporary arrays in another DO loop, thereby reducing the CPU-time of the corresponding subroutines by a factor of up to 2.89.

(d) Due to the start-up time of vector operations, the high performance

of the CRAY-1S cannot be exploited for inner loops of short vector lengths. Therefore, unrolling inner loops of vector lengths 2,3, and 4 may decrease the CPU-time considerably. Unrolling small nested loops may even enhance chaining.
Applying this guideline to the subroutine solving a block-tridiagonal linear system by the Richtmyer algorithm, resulted in a speed-up factor of 12.59. As almost 80 per cent of the total CPU-time of the original version of the blunt body code was spent in that subroutine, its modification reduced the overall computing time by a factor of almost 3.8.

(e) To compute the sum of $|\frac{\Delta \varrho}{\Delta \tau}^n|, |\frac{\Delta \varrho u}{\Delta \tau}^n|, |\frac{\Delta \varrho v}{\Delta \tau}^n|, |\frac{\Delta e}{\Delta \tau}^n|$ of all grid points in the subroutine checking the convergence, the CRAY-1 FORTRAN intrinsic function SSUM was employed. The use of SSUM and ISMAX, which determined the indices of the maximum residuals, led to a speed-up factor of 3.32 for this subroutine, and to incompatibility with standard FORTRAN. To facilitate the implementation of any program on different computers, a standardization of FORTRAN intrinsic vector functions would be desirable.

(f) Changing the dimension of the block-tridiagonal matrices from (4,4,•) to (5,5,•) and the dimension of the right hand side vector from (4,•) to (5,•), led to a speed-up factor of 2.55 for the calculation of the implicit numerical dissipation due to the elimination of the memory bank conflicts, whereas the solution of the block-tridiagonal linear system by the Richtmyer algorithm took 2 per cent more time due to the increased data transfer. The overall CPU-time was decreased by a factor of 1.06.

(g) The interdependencies among the elements of the respective vectors inhibit the vectorization of the recursions in the Richtmyer algorithm. Therefore the subroutine solving a block-tridiagonal linear system reached a computing speed of only 14.5 MFLOPS on the CRAY-1S compared with 9.5 MFLOPS, if the autovectorizer was turned off. How to remove these dependencies by modifying the Richtmyer algorithm, will be explained in the next chapter.

9.2.7. Solution of Block-Tridiagonal Linear Systems on Vector Computers

(a) Richtmyer algorithm

The solution of the block-tridiagonal linear system

(9) $\quad A\,x = f$

is to be found, where

$$A = \begin{pmatrix} A_1 & C_1 & & & \\ B_2 & A_2 & C_2 & & \\ & B_3 & A_3 & C_3 & \\ & \ddots & \ddots & \ddots & \\ & & B_{K-1} & A_{K-1} & C_{K-1} \\ & & & B_K & A_K \end{pmatrix}, \quad X = \begin{pmatrix} x_1 \\ x_2 \\ x_3 \\ \vdots \\ x_{K-1} \\ x_K \end{pmatrix}, \quad f = \begin{pmatrix} f_1 \\ f_2 \\ f_3 \\ \vdots \\ f_{K-1} \\ f_K \end{pmatrix}$$

with $\quad A_k, B_k, C_k, \quad k = 1,\ldots,K, \ \ell \times \ell$ matrices and
$x_k, f_k, \quad k = 1,\ldots,K, \ \ell$-component column vectors
(In the present case, ℓ is equal 4).

The Gauss elimination method reduces the block-tridiagonal matrix A to the product of a lower matrix L and an upper matrix U, i.e.

$$A = LU,$$

where

$$L = \begin{pmatrix} L_1 & & & & \\ B_2 & L_2 & & & \\ & B_3 & L_3 & & \\ & & \ddots & \ddots & \\ & & & B_{K-1} & L_{K-1} & \\ & & & & B_K & L_K \end{pmatrix}, \quad U = \begin{pmatrix} I & U_1 & & & \\ & I & U_2 & & \\ & & I & U_3 & \\ & & & \ddots & \ddots \\ & & & & I & U_{K-1} \\ & & & & & I \end{pmatrix}$$

and solving the triangular linear systems $Ly = f$ and $Ux = y$. Explicitly, the Richtmyer algorithm then takes the form:

$$
\begin{aligned}
& L_1 = A_1 \quad , U_1 = L_1^{-1} C_1, y_1 = L_1^{-1} f_1 \\
& L_k = A_k - B_k U_{k-1}, U_k = L_k^{-1} C_k, y_k = L_k^{-1} (f_k - B_k y_{k-1}) \ , \\
& \qquad\qquad k = 2,\dots,K-1 \\
& L_K = A_K - B_K U_{K-1} , \qquad y_K = L_K^{-1} (f_K - B_K y_{K-1}) \\
& x_K = y_K \\
& x_k = y_k - U_k x_{k+1} \ , \qquad k = K-1,\dots,1 \ .
\end{aligned} \tag{10}
$$

The $\ell \times \ell$ matrices U_k, $k = 1,\dots,K-1$, and the ℓ-component column vectors y_k, $k = 1,\dots,K$, are computed by applying the conventional Gauss elimination method to solve the ℓ linear systems

$$L_k U_k = C_k \text{ for each } k = 1,\dots,K-1,$$

and the linear system

$$L_1 y_1 = f_1 \quad \text{and}$$

$$L_k y_k = f_k - B_k y_{k-1} \text{ for each } k = 2,\dots,K.$$

As the LU decomposition of each L_k, $k = 1,\dots,K$, and the solution of the corresponding linear systems contain recursions, and as

$$U_k = (A_k - B_k U_{k-1})^{-1} C_k, \qquad k = 2,\dots,K-1, \text{ and}$$

$$y_k = L_k^{-1} (f_k - B_k y_{k-1}) \qquad k = 2,\dots,K$$

$$x_k = y_k - U_k x_{k+1}, \qquad k = K-1,\dots,1,$$

are defined recursively, the resulting dependencies preclude the vectorization of the Richtmyer algorithm.

(b) Simultaneous Richtmyer Algorithm

Amongst various possibilities, the simultaneous treatment of the Thomas algorithm, i.e. the Richtmyer algorithm for ℓ equal 1, was found in general to be the most efficient algorithm for the solution of tridiagonal linear systems on the CRAY-1S and the CYBER 205 |19|,|20| (cf. chapters 5.4 and 6.3). Carrying this result over to block-tridiagonal linear systems where ℓ is greater than one, only the simultaneous treatment of the Richtmyer algorithm will be considered here.

Since in the ξ-sweep of the Beam and Warming scheme, the block-tridiagonal linear systems for determining Δq^{*n} on lines of constant η (cf. Figure 1) are independent of each other, the dependencies stated above can be removed, if the Richtmyer algorithm is applied to these systems simultaneously by sweeping along the lines of constant ξ. Analogously, the block-tridiagonal linear systems of the η-sweep may be solved simultaneously by sweeping along the lines of constant η.

For the solution of M block-tridiagonal linear systems

$$A_m x_m = f_m \quad , \quad m = 1,\dots,M,$$

which are mutually independent, the simultaneous Richtmyer algorithm may be expressed as follows:

(11)

$$
\begin{aligned}
& k = 1 \\
& \qquad L_{k,m} = A_{k,m} && , \quad m = 1,\dots,M \; , \\
& \qquad U_{k,m} = L_{k,m}^{-1} C_{k,m} && , \quad m = 1,\dots,M \; , \\
& \qquad y_{k,m} = L_{k,m}^{-1} f_{k,m} && , \quad m = 1,\dots,M \; . \\
& k = 2,\dots,K-1 \\
& \qquad L_{k,m} = A_{k,m} - B_{k,m} U_{k-1,m} && , \quad m = 1,\dots,M \; , \\
& \qquad U_{k,m} = L_{k,m}^{-1} C_{k,m} && , \quad m = 1,\dots,M \; , \\
& \qquad y_{k,m} = L_{k,m}^{-1} (f_{k,m} - B_{k,m} y_{k-1,m}) && , \quad m = 1,\dots,M \; .
\end{aligned}
$$

$k = K$

$$L_{k,m} = A_{k,m} - B_{k,m} U_{k-1,m} \quad , \quad m = 1,\dots,M \; ,$$

$$y_{k,m} = L_{k,m}^{-1} (f_{k,m} - B_{k,m} y_{k-1,m}) \quad , \quad m = 1,\dots,M \; ,$$

$$x_{k,m} = y_{k,m} \quad , \quad m = 1,\dots,M \; .$$

$k = K-1,\dots,1$

$$x_{k,m} = y_{k,m} - U_{k,m} x_{k+1,m} \quad , \quad m = 1,\dots,M \; .$$

Each of the componentwise scalar operations in (10) becomes a vector operation with a vector length of M in (11). Compared with the Richtmyer algorithm applied to M block-tridiagonal linear systems separately, the operation count is the same for the simultaneous Richtmyer algorithm. Considering the storage requirements,

$3*K*\ell*\ell$	floating point words for A_k, B_k, C_k,	$k = 1,\dots,K$,
$K*\ell$	floating point words for f_k,	$k = 1,\dots,K$,
$(\ell*\ell)+\ell$	floating point words for the solution of the linear systems to determine U_k and y_k,	

i.e. 52xK+20 floating point words in all for ℓ equal 4, are needed for the Richtmyer algorithm. For the simultaneous Richtmyer algorithm the amount of storage required is increased by a factor of M.

Thus by the choice of the Richtmyer algorithm, the minimum number of arithmetic operations, and, by virtue of its simultaneous treatment, the maximum vector performance, are obtained at the cost of increased storage requirements. This may preclude the application of the simultaneous Richtmyer algorithm, where there is insufficient memory.

9.2.8 Advantages of Vectorization

Two versions of the blunt body code for the solution of the axisymmetric thin-layer Navier-Stokes equations by the Beam and Warming scheme (cf. Figure 6) are compared in Table 1 for the example of chapter 9.2.5. In the Richtmyer version, all of the most time consuming subroutines, except

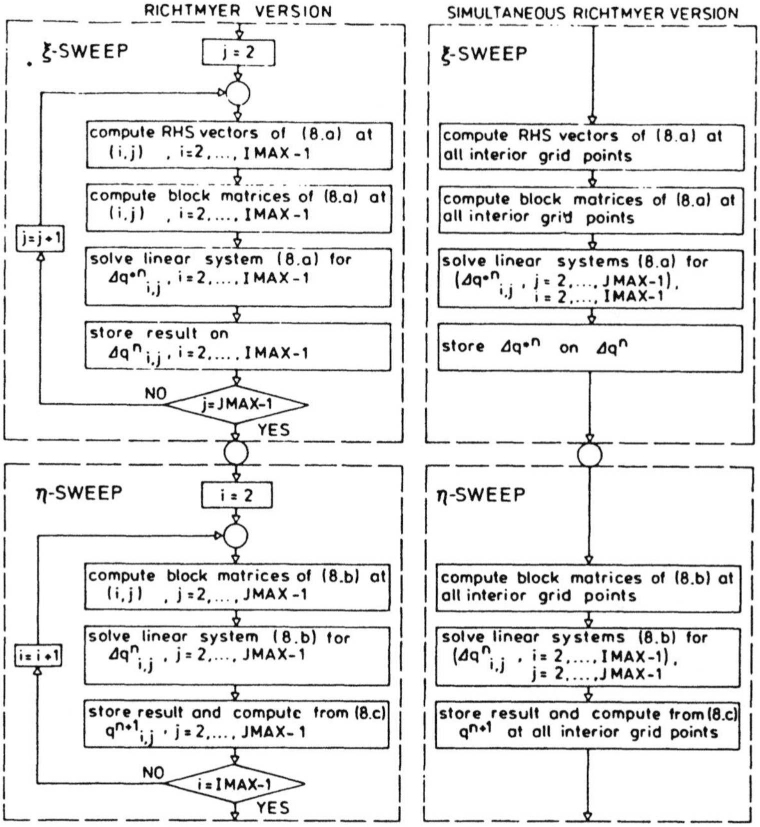

Fig. 6: Flow charts of the ξ- and η-sweeps of the Richtmyer and simultaneous Richtmyer versions (RHS = right hand side).

the one solving block-tridiagonal linear systems by the Richtmyer algorithm, are vectorized. Moreover the idea of simultaneously applying the Richtmyer algorithm is used in the simultaneous Richtmyer version.

Considering first the common subroutines of both versions, it is obvious that the computing time for input, grid generation, and initialization is almost negligible. It is interesting to realize that the grid movement, i.e. updating the positions and the metric terms of the grid points, takes only little more time than the boundary treatment, and even less than the rather costly check of convergence. No further attempt was made to speed-up the extensive output, as only a small overall improvement could be expected.

Table 1: CRAY-1S CPU-times of Richtmyer and simultaneous Richtmyer versions of the blunt body code for 400 time levels on a 26 x 31 grid.

	RICHTMYER VERSION (sec)		SIMULTANEOUS RICHTMYER VERSION (sec)	
INPUT	0.01		0.01	
GRID GENERATION	0.02		0.02	
INITIALIZATION	0.03		0.03	
GRID MOVEMENT	0.40		0.40	
ξ-SWEEP	11.88		7.59	
RHS		3.73		3.56
MATRICES		1.25		1.16
RICHTMYER		6.83		2.82
STORE		0.07		0.05
η-SWEEP	11.16		7.20	
MATRICES		4.20		4.11
RICHTMYER		6.86		3.01
STORE		0.10		0.08
BOUNDARY TREATMENT	0.29		0.29	
CONVERGENCE	0.54		0.54	
OUTPUT	1.91		1.91	
TOTAL	26.24		17.99	

The differences between the Richtmyer and the simultaneous Richtmyer versions are reflected by the execution times of the ξ- and η- sweeps. For the calculation of the block-tridiagonal matrices and the right hand sides, the simultaneous Richtmyer version takes a factor of 1.04 less time than the Richtmyer version, as the number of CALL statements is reduced and the implicit treatment of the symmetry and outflow boundary conditions becomes vectorizable. But the main reason for the reduction of the total CPU-time

by a factor of 1.46 is due to the vectorization of the block-tridiagonal system solver. Compared with the conventional Richtmyer algorithm (cf. chapter 9.2.7(a)), the simultaneous Richtmyer algorithm (cf. chapter 9.2.7(b)) attains speed-up factors of 2.42 and 2.28 for the solution of the linear systems of the ξ-sweep and the η-sweep, respectively. The corresponding computing speeds of 35 and 33 MFLOPS on the CRAY-1S for vectorlengths of 29 and 24, resp., can still be increased for longer vectors.
The loss of the vector performance becomes apparent, if the autovectorizer is turned off by the compiler option OFF = V: the computing speed of the simultaneous Richtmyer algorithm to solve the block-tridiagonal linear systems is degraded to 11.25 MFLOPS.

Table 2: CPU-times in seconds per time level and per grid point for the blunt body code.

	Richtmyer Version	Simultaneous Richtmyer Version
IBM 3081K	$101.5 * 10^{-5}$	$120.8 * 10^{-5}$
CRAY-1S	$8.1 * 10^{-5}$	$5.6 * 10^{-5}$

Considering the implementation of the blunt body code on the CYBER 205, the advantages of the simultaneous Richtmyer version will carry over. Compared with the CRAY-1S, a higher performance for forming the block-tridiagonal linear systems can be expected, because long, contiguously stored vectors can be used to take advantage of the higher peak MFLOPS rate of the CYBER 205 (cf. chapter 2.3, |21|). But for the simultaneous solution of the systems (cf. chapter 6.3 for the simultaneous treatment of the Thomas-algorithm) and also for the calculation of the systems in the Richtmyer version of the program, not all of the vectors are contiguously stored and the vector lengths are small. Therefore the high vector performance of the CYBER 205 can only be exploited to a small degree by the implicit Beam and Warming scheme.

Since longer vectors can be formed and arrays instead of vectors can be used, resp., the Beam and Warming scheme is expected to be more efficient on a vector computer like the CYBER 205 and a parallel processor like the former Illiac IV |22|, resp., for three-dimensional time-dependent problems than for two-dimensional ones, provided there is sufficient memory.

To estimate the gain of performance by the CRAY-1S, the Richtmyer and the simultaneous Richtmyer versions of the blunt body code were run on the IBM 3081K using the AUTODBL compiler option and with the CRAY-1 FORTRAN intrinsic functions ISMAX and SSUM replaced by standard FORTRAN statements. The vectorization of the original version of the program proved to be profitable on the IBM 3081K as well: The Richtmyer version led to a reduction of CPU-time by a factor of 1.37. Because of increased paging, the simultaneous Richtmyer version reached a factor of only 1.15.

Comparing the Richtmyer version on the IBM 3081K with the simultaneous Richtmyer version on the CRAY-1S (cf. Tables 2 and 3), the CPU-time per grid point and per time level was reduced from 101.5×10^{-5} sec to 5.6×10^{-5} sec, corresponding to a speed-up factor of 18.19.

As the Richtmyer algorithm cannot fully exploit the high vector performance of the CRAY-1S, there is still need for a vectorizable block-tridiagonal system solver with little storage requirements, if the available storage prohibits the application of the simultaneous Richtmyer algorithm.

Table 3: Speed-up factors of the Richtmyer Version (RV) and the Simultaneous Richtmyer Version (SRV) of the blunt body code.

RV on IBM 3081K / RV on CRAY-1S	SRV on IBM 3081K / SRV on CRAY-1S	RV on IBM 3081K / SRV on CRAY-1S
12.47	21.66	18.19

9.2.9 Conclusions

The vectorization of the implicit finite-difference scheme of Beam and Warming has been illustrated by the solution of the axisymmetric thin-layer Navier-Stokes equations to simulate supersonic flows over blunt bodies.

Solving the block-tridiagonal linear systems by the Richtmyer algorithm simultaneously instead of separately, increases the efficiency of the implicit scheme on vector processors at the cost of increased storage requirements. The concept of vectorizing the Beam and Warming scheme and the Richtmyer algorithm may be applied to the 2-D, axisymmetric, and 3-D solution of the unsteady Euler and Navier-Stokes equations and the steady hyperbolic Euler and parabolized Navier-Stokes equations by any implicit method employing solvers of mutually independent tridiagonal linear systems.

Acknowledgments

The author thanks his colleagues Prof. Dr. D. Rues for supporting this work by valuable discussions and Dr. W. Gentzsch for the introduction to vector processing. The suggestions of Prof. E. Krause, Ph.D., Aerodynamisches Institut , RWTH Aachen, are also gratefully acknowledged.

References

1. Beam R.M., Warming R.F.: An implicit factored scheme for the compressible Navier-Stokes equations. AIAA J., Vol. 16, no. 4, 1978, 393-402.
2. Kutler P., Chakravarthy S.R., Lombard C.P.: Supersonic flow over ablated nosetips using an unsteady implicit numerical procedure. AIAA Paper 78-213, 1978.
3. Peyret R., Viviand H.: Computation of viscous compressible flows based on the Navier-Stokes equations. AGARD-AG-212, 1975.
4. Baldwin B.S., Lomax H.: Thin-Layer approximation and algebraic model for separated turbulent flows. AIAA Paper 78-257, 1978.

5. Steger J.L.: Implicit finite-difference simulation of flow about arbitrary two-dimensional geometries. AIAA J., Vol. 16, no. 7, 1978, 679-686.
6. Pulliam T.H.,Steger J.L.: On implicit finite-difference simulations of 3D flow. AIAA Paper 78-10, 1978.
7. Roberts G.O.: Computational meshes for boundary-layer problems. Proc. 2nd Int. Conf. Numer. Meth. in Fluid Dynamics, Lecture Notes in Physics 8, Springer-Verlag, New York 1971.
8. Billig F.S.: Shock-wave shapes around spherical- and cylindrical nosed bodies. J. Spacecraft, Vol. 4, no. 6, 1967, 822-823.
9. Hayes W.D., Probstein R.F.: Hypersonic flow theory. 2nd ed., Vol. I, Inviscid Flows, Academic Press, New York 1966.
10. Chakravarthy S.R.,Anderson D.A., Salas M.D.: The split-coefficient matrix method for hyperbolic systems of gas dynamic equations. AIAA Paper 80-0268, 1980.
11. Rues D.: Der Einfluss einfallender Stosswellen auf ebene Ueberschallstroemungen um stumpfe Koerper. DLR-FB 72-68, 1972.
12. Kutler P., Pedelty J.A., Pulliam T.H.: Supersonic flow over three-dimensional ablated nosetips using on unsteady implicit numerical procedure. AIAA Paper 80-0063, 1980.
13. Viviand H., Ghazzi W.: Numerical solution of the Navier-Stokes equations at high Reynolds numbers with application to the blunt body problem. Proc. 5th Int. Conf. Numer. Meth. in Fluid Dynamics, Lecture Notes in Physics 59, Springer-Verlag, Berlin 1976.
14. Hsieh T.:Heat transfer calculation for hypersonic flows over blunt noses using an unsteady implicit scheme.Proc.2nd Nat.Symp.Numer. Meth. in Heat Transfer, Hemisphere Publ., Washington DC, 1982.
15. Hsieh T.: Calculation of viscous hypersonic flow over a severely indented nosetip. AIAA Paper 83-0226, 1983.
16. Sydow P.J.:Optimization guide.CRAY Computer Systems Technical Note SN-0220, Revision A, CRAY RESEARCH INC., May 1982.
17. CRAY-1 Computer systems FORTRAN (CFT) reference manual SR-0009, Revision I, CRAY RESEARCH INC., Nov. 1982.
18. Isaacson E.,Keller H.B.: Analysis of numerical methods. John Wiley and Sons, New York 1966.
19. Hockney R.W., Jesshope C.R.: Parallel computers. Adam Hilger, Bristol 1981.
20. Gentzsch W.:Recursion algorithms on vector computers.Proc.1st Int.

Conf. on Vector and Parallel Computing in Scientific Applications, Paris 1983, 79-86.

21. Rizzi A.: Vector coding the finite volume procedure for the CYBER 205. VKI Lecture Series 1983-04, von Karman Institute, Rode Saint Genese 1983.
22. Pulliam T.H.,Lomax H.: Simulation of three-dimensional compressible viscous flow on the Illiac IV computer. AIAA Paper 79-0206, 1979.
23. Mueller B.: Calculation of axisymmetric laminar supersonic flow over blunt bodies. DFVLR report, Goettingen 1984.

9.3 Vectorization of an Implicit Finite Difference Method for the Solution of the Laminar Boundary-Layer Equations

Dieter Schwamborn, DFVLR

9.3.1 Introduction

As was shown by L. Prandtl in 1904 |1|, the flow around a body can be subdivided into two constituents; one thin layer in the vicinity of the surface, where friction is important, and the remaining part of the flow field which can be treated as inviscid. Since then, the theory of this thin viscous layer, the boundary layer, has developed into a very important branch of fluid mechanics; only by using boundary-layer theory, it has become possible to explain various flow phenomena, and to treat flows around bodies, including viscous effects, both theoretically and numerically.

A major disadvantage of boundary-layer theory, however, is that it is no longer valid in regions of the surface where the boundary-layer thickness is not small compared with the smallest local radius of curvature of the surface (first order boundary-layer theory), or where separation occurs. To overcome these problems, it is necessary to solve the Navier-Stokes equations, but despite much progress towards their solution for a variety of flow configurations, the boundary-layer calculations have lost neither their usefulness nor their attraction. This is especially true for three-dimensional flow fields and flow at high Reynolds numbers, where Navier-Stokes codes would require more storage for sufficient accuracy, and more computation time, than practicable on contemporary computers.

The latter argument is, however, being severely eroded with the recent introduction of the so-called supercomputers, (or at least their next generation) which can significantly reduce the calculation time for a given problem (provided that the "right" algorithm is used for the solution); storage capacity, however, continues to be a problem.

In the present chapter, we shall deal with an algorithm for the numerical solution of the boundary-layer equations (by the boundary layer equations we always mean the complete set of differential equations for continuity, momentum and energy) and its implementation on a CRAY-1 vector computer. In section 9.3.2, we begin with a presentation of the boundary-layer equations for an appropriate coordinate system using tensorial concepts. The boundary-layer equations are solved using a finite difference method, which shall be discussed in the third section where a simplified flow chart of the algorithm is given. The method is based on |2,3,4|. The reader interested in other methods is referred to the reviews of Blottner |5,6|. The fourth section deals with the changes made in the boundary-layer code in order to vectorize the most time consuming parts. In the same section, the resulting computation time on the CRAY-1S is discussed for the example of the calculation of a part of the boundary layer on a winglike spheroid based on |7| and a comparison with the IBM 3081K performance is made.

9.3.2 The Coordinate System and the Boundary-Layer Equations

For boundary-layer calculations it is necessary to use a coordinate system in which a family of coordinate lines (e.g. x^1=const., x^2= const.) is rectilinear and perpendicular to the surface, represented by $x^3=0$. Such coordinate systems are known as surface-oriented, locally monoclinic, coordinate systems |8,9|. In general, the two other families of coordinate lines are nonorthogonal and curvilinear. For boundary-layer calculations on wings, the lines x^1=const., $x^3=0$ and x^2=const., $x^3=0$ are often lines of constant chord and constant span respectively. An advantage of the locally monoclinic coordinates is that all metric properties of the coordinate system can be determined using only the metric properties of the surface. The changes in the coordinate system away from the surface are then represented, by the use of the so-called "shifters". These are only needed, however, where first -order boundary-layer theory is no longer applicable. The interested reader is referred to |8,9|.

We describe the surface of a body in cartesian coordinates $x^{i'}$ ($i'=1,2, 3$) by use of the two Gaussian surface-parameters x^{α} ($\alpha=1,2$) of the locally monoclinic, coordinate system (x^1,x^2,x^3), i.e.:

(1) $\quad x^{i'} = x^{i'}\,(x^1,x^2) \qquad i = 1,2,3.$

This transformation is considered to be unique and enables us to determine all metric properties needed in the following. We begin by presenting the first-order, steady-state, boundary-layer equations in their dimensionless form: (Time-dependent calculations will not be considered here).

Continuity equation:

$$(k_{01}\rho v^1)_{,1} + (k_{01}\rho v^2)_{,2} + (k_{01}\rho v^3)_{,3} = 0 \quad , \tag{2}$$

Momentum equation for x^1-direction

$$\rho[v^1 v^1_{,1} + v^2 v^1_{,2} + v^3 v^1_{,3} + k_{11}(v^1)^2 + k_{12}v^1 v^2 + k_{13}(v^2)^2] =$$

$$= k_{14}p_{,1} + k_{15}p_{,2} + (\mu v^1_{,3})_{,3} = \tag{3}$$

$$= \rho_e[v^1_e v^1_{e,1} + v^2_e v^1_{e,2} + k_{11}(v^1_e)^2 + k_{12}v^1_e v^2_e + k_{13}(v^2_e)^2] + (\mu v^1_{,3})_{,3} \quad ,$$

Momentum equation for x^2-direction

$$\rho[v^1 v^2_{,1} + v^2 v^2_{,2} + v^3 v^2_{,3} + k_{21}(v^1)^2 + k_{22}v^1 v^2 + k_{23}(v^2)^2] =$$

$$= k_{24}p_{,1} + k_{25}p_{,2} + (\mu v^2_{,3})_{,3} = \tag{4}$$

$$= \rho_e[v^1_e v^2_{e,1} + v^2_e v^2_{e,2} + k_{21}(v^1_e)^2 + k_{22}v^1_e v^2_e + k_{23}(v^2_e)^2] + (\mu v^2_{,3})_{,3} \quad ,$$

Energy equation

$$c_p\rho[v^1 T_{,1} + v^2 T_{,2} + v^3 T_{,3}] = \frac{1}{Pr_{ref}}(KT_{,3})_{,3} + E_{ref}\{(v^1 p_{,1} + v^2 p_{,2}) +$$

$$+ \mu[k_{41}(v^1_{,3})^2 + k_{42}v^1_{,3}v^2_{,3} + k_{43}(v^2_{,3})^2]\} \quad . \tag{5}$$

All quantities in these equations are non-dimensionalized by using the appropriate reference values. For the pressure, $\rho_{ref}\, U^2_{ref}$ is used as reference value and, furthermore, the velocity and the coordinate normal to the surface v^3 and x^3 are stretched by multiplying by the square root of the reference Reynolds number. The contravariant velocities v^α are related to the physical velocities $\overset{*}{v}^\alpha$ by

$$\overset{*}{v}^\alpha = v^\alpha \sqrt{a_{(\alpha\alpha)}} \quad ,$$

where $a_{(\alpha\alpha)}$ is a covariant component of the metric-tensor of the surface

$$a_{\alpha\beta} = \sum_{K=1}^{3} \beta_{\alpha}^{K'} \beta_{\beta}^{K'} \qquad \alpha = 1,2 \quad .$$

The properties $\beta_j^{K'}$ are elements of the transformation matrix $(\beta_j^{K'})$ and can be determined from equation (1) with

$$\beta_{\alpha}^{K'} = \frac{\partial x^{K'}}{\partial x^{\alpha}} \qquad \alpha = 1,2$$

and $\beta_3^{K'}$ from the conditions of orthonormality

$$\sum_{K=1}^{3} \beta_{\alpha}^{K'} \beta_{3}^{K'} = 0 \qquad \alpha = 1,2$$

$$\sum_{K=1}^{3} (\beta_3^{K})^2 = 1 \quad .$$

The elements of the transformation matrix and its inverse, the Jacobian of the transformation, serve also to transform contravariant components of a vector into Cartesian ones, and vice versa:

$$v^{K'} = \beta_j^{K'} v^j$$

$$v^i = \beta_{K'}^{i} v^{K'} \quad .$$

The metric properties k_{ij} used in the boundary-layer equations (2-5) are determined from the elements of the metric tensor of the surface:

$$k_{o1} = [a_{11}a_{22} - (a_{12})^2]^{1/2} ,$$

$$k_{11} = \frac{1}{2a} [a_{22}a_{11'1} - a_{12}(2a_{12'1} - a_{11'2})] ,$$

$$k_{12} = \frac{1}{2} [a_{22}a_{11'2} - a_{12}a_{22'1}] ,$$

$$k_{13} = \frac{1}{2a} [a_{22}(2a_{12'2} - a_{22'1}) - a_{12}a_{22'2}] ,$$

$$k_{14} = - \frac{a_{22}}{a} ,$$

$$k_{15} = \frac{a_{12}}{a} ,$$

$$k_{21} = \frac{1}{2a} [a_{11}(2a_{12'1} - a_{11'2}) - a_{12}a_{11'1}] ,$$

$$k_{22} = \frac{1}{a} [a_{11}a_{22,1} - a_{12}a_{11,2}] ,$$

$$k_{23} = \frac{1}{2a} [a_{11}a_{22,2} - a_{12}(2a_{12,2} - a_{22,1})] ,$$

$$k_{24} = \frac{a_{12}}{a} = k_{15} ,$$

$$k_{25} = - \frac{a_{11}}{a} ,$$

$$k_{41} = a_{11} ,$$

$$k_{42} = 2a_{12} ,$$

$$k_{43} = a_{22} ,$$

$$a = (k_{01})^2 .$$

Besides the boundary-layer equations, (2-5) we require the equation of state in its dimensionless form:

(6) $$p = \frac{\rho T}{\gamma_{ref} M^2_{ref}}$$

and, furthermore, the laws of viscosity and thermal conductivity as expressed by:

$$\mu = \mu(T) \quad \text{and} \quad K = K(T) .$$
(7)

The boundary-layer conditions for the differential equations are at the wall $x^3 = 0$:

$$v^1 = v^2 = 0 , \quad v^3 = v^3_w \, (=0) ,$$
(8a)
$$T = T_w \quad \text{or} \quad T_{,3} = T_{w,3}$$

and at the outer edge of the boundary layer $x^3 = \delta$:

(8b) $$v^1 = v^1_e(x^1, x^2) , \quad v^2 = v^2_e(x^1, x^2) , \quad T = T_e(x^1, x^2) .$$

Moreover initial conditions are required for any three-dimensional

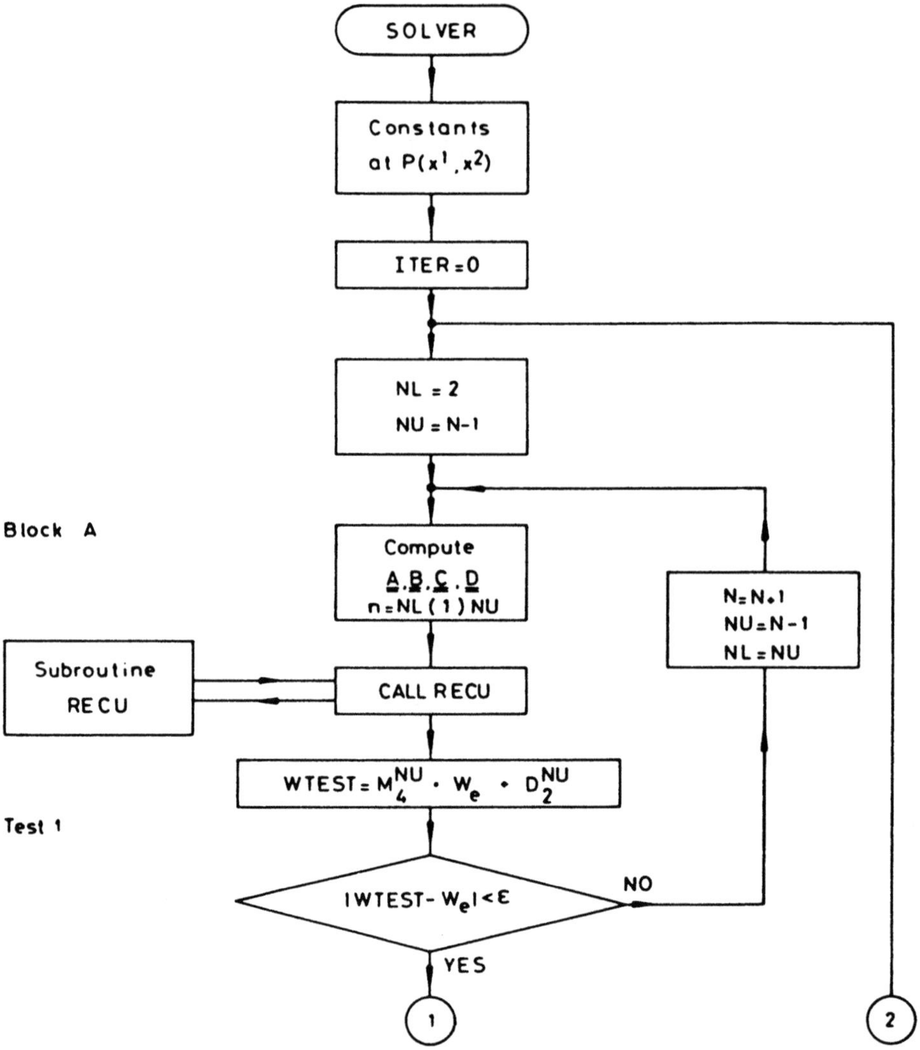

Fig. 1a: Flow chart of the boundary-layer code, part 1.

boundary-layer calculation in at least one plane normal to the wall. These initial conditions may be obtained in different ways depending on the given configuration; e.g. by use of a symmetry plane solution |7,9| which will usually include a stagnation-point solution or a solution using the "locally infinite swept wing" concept |9|.

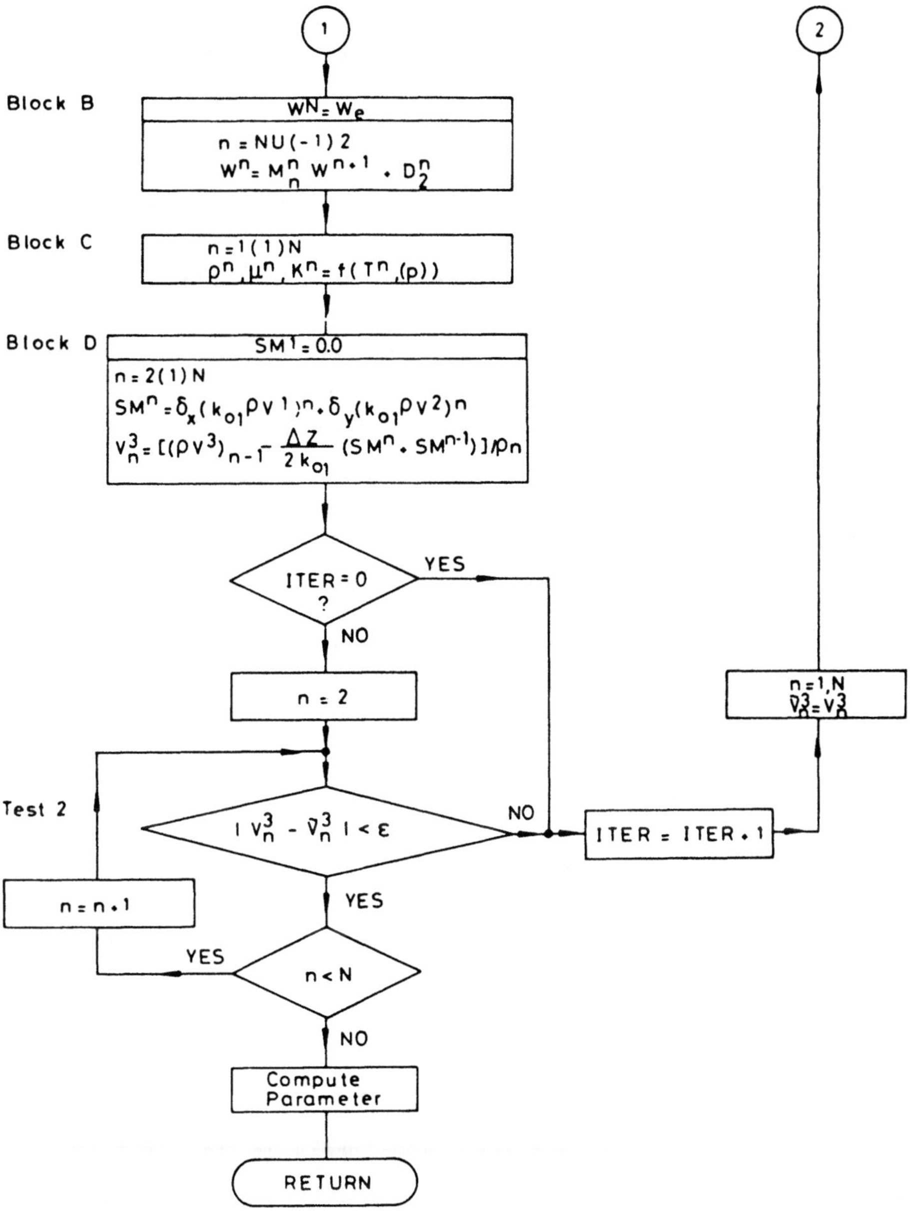

Fig. 1b: Flow chart of the boundary-layer code, part 2.

From the solution of the boundary-layer equations one can obtain boundary-layer parameters such as wall shear stress, wall heat flux and different kinds of displacement thicknesses which will not be discussed here for sake of brevity. The interested reader is therefore referred to |9|.

9.3.3 Finite Difference Solution of the Boundary-Layer Equations

In this chapter, we shall be using the second-order accurate method of Krause |2| (for two-dimensional problems) and Krause et al. |3,4| (for three-dimensional flow) to solve the boundary-layer equations for a given geometrical configuration and boundary conditions. For stability reasons, this method uses a formulation which is implicit in the x^3- direction (normal to the surface).

Irrespective of the special case of the boundary-layer equations |9| and of the difference scheme |4,7| used for the discretization of the system of partial differential equations (PDE) for momentum and energy (3-5), the discretization can be considered as follows. First the PDE's are discretized in the surface-tangential directions yielding a system of ordinary differential equations of the form

$$(9) \qquad A\,W + B\,W' + C\,W'' + D = 0$$

where W represents the unknown solution vector $(v^1, v^2, T)^T$ at some position $P(x^1, x^2)$, and the prime denots the derivative with respect to the surface normal direction (x^3). Because of the non-linearity of the equations of motion, the matrices A,B,C and vector D depend not only on known data, but also on the solution at $P(x^1, x^2)$ itself.

The difference scheme used for special cases of the boundary-layer equations, (as e.g. for quasi-one-dimensional stagnation-point flow, or quasi-two-dimensional symmetry plane flow |7,9|), influences only the elements of the matrices and the vector in (9), but not the form of this equation (We do not give information about difference schemes for boundary-layer calculations, but refer the reader e.g. to |3,7|). Thus from this point on one subroutine can serve for the solution of the boundary-layer equations (except the continuity equation), independent of the flow. We now discretize equation (9) using the central 3-point formula, yielding block-tridiagonal systems of -2 difference equations, where N is the unknown number of gridpoints from the wall to the edge of the boundary layer:

$$M_1^n W^{n+1} + M_2^n W^n + M_3^n W^{n-1} + D_1^n = 0 \qquad 2 \le n \le N-1 . \tag{10}$$

This system is closed by the bounda conditions (8):

$$x^3 = 0 : \quad W^{n=1} = W_{Wall}(x^1, x^2)$$
(11)
$$x^3 = \delta : \quad W^{n=N} = W_e(x^1, x^2)$$

To give a better impression of the algorithm presented here, and the changes to be made in the next chapter, a simplified flow chart of the main part of a boundary-layer code is shown in Fig. 1. Every code for the calculation of boundary layers, can be written in such a way that different subroutines are used for any position on the surface where different equations (for quasi-one-, quasi-two-, or three-dimensional flow) or different difference schemes are used. Such a subroutine called SOLVER is represented by our flow chart. We assume that all data needed in the routine have been stored in the appropriate places by the calling routine beforehand. SOLVER computes firstly all constants and arrays not changing during the calculation at one position $P(x^1,x^2)$ (as e.g. the pressure terms or normal derivatives at neighbouring positions). The iteration count (ITER) is set to zero and the first loop (IF Loop) is entered. Here the elements of the matrices A,B,C and vectors D (equ. 9) are computed from n = 2 to n = N-1. For the unknown data in these elements an estimate from neighbouring points is used at first, as is done for N. When the loop is terminated the subroutine RECU (Fig. 2) is called, which is the same for all subroutines similar to SOLVER. This routine consists of one DO-loop which starts with n = NL and ends with n = NU, where now NL = 2 and NU = N-1. At first the elements of the matrices in (10) are computed from the elements of A,B,C and the stepsizes in direction normal to the surface (Block 1).

Using the boundary condition at the wall, equation (10) can be transformed into

(12) $$W^n = M_4^n W^{n+1} + D_2^n \qquad , \quad 2 \le n \le N-1$$

where the matrices M_4 and vectors D_2 can be calculated successively from the wall outwards by:

(13)
$$M_K^n = -(M_2^n + M_3^n \cdot M_4^{n-1})^{-1} \cdot M_1^n$$
$$D_2^n = -(M_2^n + M_3^n \cdot M_4^{n-1})^{-1} \cdot (D_1^n + M_3^n \cdot D_2^{n-1}) \qquad 2 \le n \le N-1 .$$

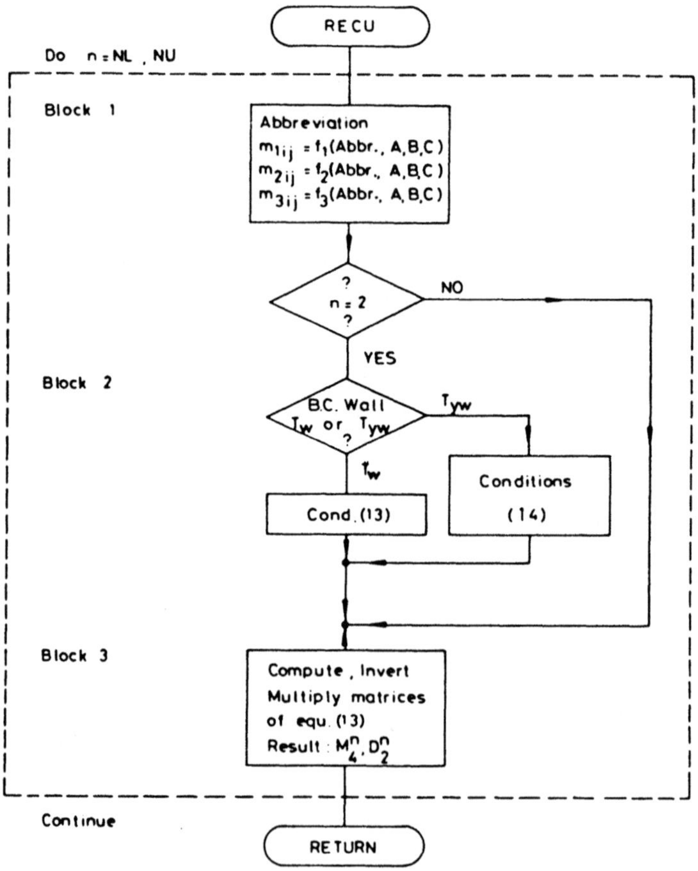

Fig 2: Flow chart of the scalar subroutine RECU.

This is carried out in Block 3 of RECU. To consider the boundary condition at the wall, we have to distinguish the case of given wall temperature T_W with:

(14) $\quad M_4^{n=1} = 0 \quad ; \quad D_2^{n=1} = W^{n=1} = (0,0,T_W)^T$

from the case of a given gradient of the temperature at the wall $T_{,3W}$, with:

$$M_4^{n=1} = 4/3\, S \quad ,$$

$$D_2^{n=1} = (0.,0.,\tfrac{2}{3}\Delta x^3 T_{,3}\, w) \quad , \qquad S = \begin{pmatrix} 0 & 0 & 0 \\ 0 & 0 & 0 \\ 0 & 0 & 1 \end{pmatrix} ,$$

$$\tilde{M}_1^{n=2} = M_1^{n=2} - M_3^{n=2} \cdot S/3 \quad ,$$

respectively, where a little modification is required in the latter case using $\tilde{M}_1$ instead of M_1 in (13), if $n = 2$. This handling of the boundary conditions is performed in Block 2 of RECU.

When the DO-loop is terminated, control is returned to SOLVER. Since the boundary-layer thickness is unknown, a priori the index n (and N) has to be incremented until the boundary condition (11) for the outer edge of the boundary layer is fulfilled within a small error ε (TEST 1):

$$|W^{n=N-1} - W_e| < \varepsilon \quad ,$$

where, by (12),

$$W^{n=N-1} = M^{n=N-1}\, W_e + D_2^{n=N-1} .$$

If N has been incremented, Block A and routine RECU are executed once again; this time with $n = NL = NU = N-1$.

If TEST 1 is fulfilled, (12) is used to calculate the solution vectors W^n from $n = N-1$ to $n = 2$ (Block B) and the density, viscosity and thermal conductivity are computed from the appropriate equations (6), (7) in Block C.

Then the continuity equation is used to obtain the normal-velocity components v^3. This normal velocity is not calculated at the same grid points as all other variables, but at the center points of the discretization for the x^1 and x^2 directions. Thus the v^3 profile is obtained without use of other v^3 profiles by simply integrating the discretized continuity equation in x^3 direction (Block D).

We have now to remember that we used estimates for the unknowns in the elements of the A,B,C and D of equation (9). To prevent a loss in accuracy, we have to recalculate these elements using the recently obtained solution for the unknowns and repeat the whole calculation. Thus an iteration process results which is terminated if the v^3- profiles of two successive iterations differ from each other by less than a certain error bound (TEST 2):

$$|{}^{\nu}v^3 - {}^{\nu-1}v^3| < \varepsilon \qquad 2 \leq n \leq N \qquad (\nu \,\hat{=}\, \mathrm{ITER}) \; .$$

Then the routine calculates all boundary-layer parameters needed and the solution proceeds to the next mesh position $P(x^1,x^2)$ where the solution is unknown and the same procedure (SOLVER) is repeated. Not indicated in the flow chart (Fig. 1) is the action to be taken if the number of points N normal to the wall or the number of iterations exceed some prescribed limit. In this case an error message is issued, the values at the last iteration are printed, and the program execution is interrupted.

9.3.4 Implementation of a Boundary-Layer Code on the CRAY-1S

The present chapter deals with the changes to be made to the FORTRAN program for boundary-layer calculations in order to maximize the degree of vectorization. We shall discuss only those changes necessary to reduce the calculation time on the CRAY-1S. It is found that, where these changes are permissible for the IBM 3081K, they have little or no impact.To gauge their effect, a boundary-layer code for compressible flow based on |8| is used to calculate the laminar boundary layer on the surface of a winglike spheroid at a Mach-number of 0.3. The CPU-time for the computation of a distinct part of this three-dimensional boundary layer is used as a yard-stick for measuring the calculation time.

For the existing code without any changes it took 414 CPU seconds on the IBM 3081K and 50.4 seconds on the CRAY-1S to perform the calculation. The computation time on the CRAY did not change significantly when the autovectorization facility was de-activated. This meant that the algorithm was initially essentially serial. It seemed then, somewhat surprising that there was a factor of 8 in the times of the IBM 3081K to the CRAY, since the scalar processor of the CRAY is not 8 times faster than that of the IBM.
This is largely attributable to the fact that there is much unformatted input and output of long data strings in the code, which is very fast on the CRAY compared to the IBM (see also |11|). We will return to the question of I/O later.

As mentioned in section 3.1, one can obtain a so-called "Flow Trace" for a calculation on the CRAY. In the case at hand, it is found by using this flow-trace option that 98 per cent of the CPU-time is spent in only two subroutines, one of these being the routine RECU described in the preceding chapter. The other routine called FO3D contains two SOLVERs - based on two, different, difference schemes - for three-dimensional flow as discussed above, as well as the data handling (mainly unformatted I/O) and management necessary between the solution steps at different positions $P(x^1,x^2)$. Since all other routines contribute only very little to the computation time, changes are only made in these two routines. For a better comparison of the changes in CPU time resulting from the improvements discussed in the following, the CPU time for the two routines RECU and FO3D as well as for the complete code is given in Table 1.

The flow trace, as mentioned, indicates which subroutine is the most time-consuming one, and should therefore be improved first. Which parts of the subroutine should be changed may be determined by the messages appended to the subroutine listing. Here the CRAY-compiler supplies a list of how it blocked the routine, where it produces vector loops or short (vector) loops and when a DO loop is not vectorized due to some "vector dependency". The routine RECU has a single DO loop in which such a dependency exists due to the recurrency (equ. 13) in the third block (Fig. 2). This DO-loop can, however, be split into two, one for the first block and the second for the third one. The elements of the matrices M_1,M_2 and M_3 then have to be stored for the use in the second loop, thus increasing the amount of storage requirement; but at the same time allowing the first loop to be vectorized, and so improving performance. The additional storage is 27 times N_{max} words where N_{max} is the maximum number of points in the boundary layer allowed normal to the wall, and all arrays for the elements of A,B,C or for the flow variables are of this size. After these changes, the CPU time is reduced by about 25 per cent for RECU or about 11 per cent in total (2. in Table 1).

The next change is made in FO3D where the greatest portion of the CPU-time is now spent. Here the elements of the matrices A,B,C and vectors D for all n are computed in block A (Fig. 1) using an IF-loop.

Table 1: CPU times (in seconds) for RECU, FO3D and the complete code.

STEPS OF VECTORIZATION	RECU	FO3D	Total
1. Serial version of code on CRAY			
without autovectorization	24.80	25.69	52.10
with autovectorization	24.80	24.02	50.40
2. Split DO loop in RECU into 2 loops (27 arrays more storage) 1. loop vectorizes	18.71	24.16	44.91
3. DO loop instead of IF loop in Block A of SOLVERs (causes vectorization)	18.71	15.12	35.37
4. Split DO loop in Block D into 3 loops 1 array more storage; 1.+2. loop vectorize	18.71	13.83	34.05
5. Increase loop length in Block 1 nine times	18.23	13.83	33.57
6. 2. Short loop in RECU by splitting into cases: NL=2=NU and NL=NU	17.84	13.83	33.18
7. Buffer I/O in FO3D	17.84	11.23	30.58
8. Version 1.(above) on IBM 3081K	/	/	414.0
9. Version 4.(above) with splitting like in 6.(above) on IBM 3081K	/	/	411.0

Since a DO-loop is nothing other than a special formulation of an IF-loop, the execution time on serial computers is usually not seriously affected by using an IF-loop instead of a DO-loop or vice versa. The autovectorizer of the CRAY is, however, only capable of vectorizing DO-loops (see also |11|). Thus changing the IF-loop in block A into a DO-loop decreases the CPU time for FO3D about 37 per cent and for the whole computation about another 21 per cent (3. in Table 1).

The list of the compiler blocks for FO3D shows that there is a vector dependency in the DO loop for the integration of the continuity equation (Block D in Fig. 1) as $v^3(n)$ depends on $v^3(n-1)$. Splitting this loop into three results in two vectorizable loops, where the discretized surface-tangential derivatives of equation (2) are computed for all n in the first one. The third non-vectorizable loop consists of only one statement of the form

$$v^3(n) = v^3(n-1) * f(n-1) - g(n-1) ,$$

where f(n-1) and g(n-1) are calculated from the surface-tangential derivatives, the stepsizes and the density in the second (vectorizable) loop. For a better understanding, the old and the new formulation of this part of the code are given in the following, where U,V,W are used instead of v^1,v^2,v^3, RHO is ρ and RK01 is k_{01}. The last cipher of the arrays specifies their position in the difference scheme (not given here) and the first definition of SM is half the sum of the surface-tangential derivatives of equation (2):

```
C+++++++++++++++++++++++++++++++++++++++++++++++++++++++++++++++++++++++
C*    CONTINUITY EQUATION
C                                         OLD VERSION
      SM(1)=0.0
      RR1=RK011*RHO1(1)
      RR4=RK014*RHO4(1)
      RMA=(RR1+RR4)*0.5
      DO 600 K=2,N
      RR1=RHO1(K)*RK011
      RR2=RK012*RHO2(K)
      RR3=RK013*RHO3(K)
      RR4=RHO4(K)*RK014
      RM=(RR1+RR4)*0.5
      SM(K)=(RR4*U4(K)-RR3*U3(K)+RR2*U2(K)-RR1*U1(K))*ED4DX
     &                +(RR4*V4(K)+RR3*V3(K)-RR2*V2(K)-RR1*V1(K))*ED4DY
      W(K)=(W(K-1)*RMA-DZ(K-1)*(SM(K)+SM(K-1)))/RM
 600  RMA=RM
```

```
C++++++++++++++++++++++++++++++++++++++++++++++++++++++++++++++++++++++
C*    CONTINUITY EQUATION
C                                           NEW VERSION
      SM(1)=0.
      RR1=RK011*RH01(1)
      RR4=RK014*RH04(1)
      RM(1)=(RR1+RR4)*0.5
      DO 600 K=2,N
      RR1=RK011*RH01(K)
      RR2=RK012*RH02(K)
      RR3=RK013*RH03(K)
      RR4=RK014*RH04(K)
      RM(K)=(RR1+RR4)*0.5
 600  SM(K)=(RR4*U4(K)-RR3*U3(K)+RR2*U2(K)-RR1*U1(K))*ED4DX
     &                  +(RR4*V4(K)+RR3*V3(K)-RR2*V2(K)-RR1*V1(K))*ED4DY
      KU=N-1
      DO 602 K=1,KU
      SM(K)=(SM(K)+SM(K+1))*DZ(K)
      R=1/RM(K+1)
      RM(K)=RM(K)*R
 602  SM(K)=SM(K)*R
      DO 603 K=2,N
 603  W(K)=W(K-1)*RM(K-1)-SM(K-1)
```

From the above one can see that only one array (RM) more is needed in the new form. The improvement in time is 1.3 seconds (4.in Table 1).

Since most of the CPU time is now spent in routine RECU, we try to make further improvements there. The dependency of the elements of the matrices M_1,M_2, and M_3 on those of A,B and C is of the form

$$m(n) = f(n)*a(n) + g(n)*b(n) + h(n)*c(n) ,$$

where f,g and h depend on the stepsize normal to the wall, but are equal for one matrix M_i and constant n. (For constant stepsize, f, g, h are independent of n). Instead of writing 9 statements for each 3 x 3 matrix , one for each element m_{ij}, in the DO-loop over n in block 1 (Fig. 2), we can write only one statement for each M, if we change the index of the DO-loop and the form of the arrays used for the matrix elements. The 9 arrays of dimension N_{max} for each matrix are now changed to one three-dimensional array of dimensions (3,3,N_{max}), so that the 9 elements of each matrix for each n are now contiguously stored in memory. The DO loop index is changed and varies from (9*NL-8) to (9*NU) instead from NL to NU. The main differences between old and new formulation of routine RECU are given in the following:

```
C
C*    COEFFICIENTS IN DIFFERENCE EQU.S;  M1*F(K+1)+M2*F(K)+M3*F(K-1)+DE=0
C
C
C OLD VERSION
C
      COMMON                  A11(200),A12(200),A13(200)
     &                       ,A21(200),A22(200),A23(200)
     &                       ,A31(200),A32(200),A33(200)
     &                             ,B11(200),B12(200),B13(200)
     &                             ,B21(200),B22(200),B23(200)
     &                             ,B31(200),B32(200),B33(200)
     &                                  ,C11(200),C12(200),C13(200)
     &                                  ,C21(200),C22(200),C23(200)
     &                                  ,C31(200),C32(200),C33(200)
     &                                        ,DE1(200),DE2(200),DE3(200)
     &      ,H1RZ(200),H2RZ(200),HRZ(200),H2M1(200),DZ(200)
      REAL                    M111(200),M112(200),M113(200)
     &                       ,M121(200),M122(200),M123(200)
     &                       ,M131(200),M132(200),M133(200)
     &                             ,M211(200),M212(200),M213(200)
     &                             ,M221(200),M222(200),M223(200)
     &                             ,M231(200),M232(200),M233(200)
     &                                  ,M311(200),M312(200),M313(200)
     &                                  ,M321(200),M322(200),M323(200)
     &                                  ,M331(200),M332(200),M333(200)
C
      DO 5 K=NL,NU
      KM=K-1
      M111(K)=H1RZ(K)*(DZ(KM)*B11(K)+C11(K))
      M112(K)=H1RZ(K)*(DZ(KM)*B12(K)+C12(K))
      M113(K)=H1RZ(K)*(DZ(KM)*B13(K)+C13(K))
      M121(K)=H1RZ(K)*(DZ(KM)*B21(K)+C21(K))
      M122(K)=H1RZ(K)*(DZ(KM)*B22(K)+C22(K))
      M123(K)=H1RZ(K)*(DZ(KM)*B23(K)+C23(K))
      M131(K)=H1RZ(K)*(DZ(KM)*B31(K)+C31(K))
      M132(K)=H1RZ(K)*(DZ(KM)*B32(K)+C32(K))
      M133(K)=H1RZ(K)*(DZ(KM)*B33(K)+C33(K))
      M211(K)=A11(K)+HRZ(K)*(H2M1(K)*B11(K)-C11(K))
      M212(K)=A12(K)+HRZ(K)*(H2M1(K)*B12(K)-C12(K))
      M213(K)=A13(K)+HRZ(K)*(H2M1(K)*B13(K)-C13(K))
      M221(K)=A21(K)+HRZ(K)*(H2M1(K)*B21(K)-C21(K))
      M222(K)=A22(K)+HRZ(K)*(H2M1(K)*B22(K)-C22(K))
      M223(K)=A23(K)+HRZ(K)*(H2M1(K)*B23(K)-C23(K))
      M231(K)=A31(K)+HRZ(K)*(H2M1(K)*B31(K)-C31(K))
      M232(K)=A32(K)+HRZ(K)*(H2M1(K)*B32(K)-C32(K))
      M233(K)=A33(K)+HRZ(K)*(H2M1(K)*B33(K)-C33(K))
      M311(K)=H2RZ(K)*(-DZ(K)*B11(K)+C11(K))
      M312(K)=H2RZ(K)*(-DZ(K)*B12(K)+C12(K))
      M313(K)=H2RZ(K)*(-DZ(K)*B13(K)+C13(K))
      M321(K)=H2RZ(K)*(-DZ(K)*B21(K)+C21(K))
      M322(K)=H2RZ(K)*(-DZ(K)*B22(K)+C22(K))
      M323(K)=H2RZ(K)*(-DZ(K)*B23(K)+C23(K))
      M331(K)=H2RZ(K)*(-DZ(K)*B31(K)+C31(K))
      M332(K)=H2RZ(K)*(-DZ(K)*B32(K)+C32(K))
    5 M333(K)=H2RZ(K)*(-DZ(K)*B33(K)+C33(K))
```

```
C NEW VERSION
C
      COMMON              A(3,3,200)
     &                      ,B(3,3,200)
     &                          ,C(3,3,200)
     &                              ,DE1(200),DE2(200),DE3(200)
     &     H1RZV(1800),H2RZV(1800),HRZV(1800),H2M1V(1800),DZV(1800)
      REAL              M1(3,3,200)
     &                     ,M2(3,3,200)
     &                         ,M3(3,3,200)

      NUV=NU*9
      NLV=NL*9-8
      DO 5 L=NLV,NUV
      M1(L)=H1RZV(L)*(DZV(L-9)*B(L)+C(L))
      M2(L)=A(L)+HRZV(L)*(H2M1V(L)*B(L)-C(L))
    5 M3(L)=H2RZV(L)*(-DZV(L)*B(L)+C(L))
```

The only disadvantage of the new version is that more storage is required for the functions of the stepsize (H1RZV to DZV in the COMMON Block) since the new arrays are nine times as large as the old ones (H1RZ to DZ). If we denote one of the old arrays by q and its new version by qV we have for the contents of the new array:

$$qV(9 \cdot n-i) = q(n) \qquad i = 0,1,\ldots,8, \qquad \text{for all } n .$$

As can be seen from Table 1 (No. 5), only minimal improvements result (3 per cent for RECU), whereas the demand on storage capacity is considerably increased. (40 * N_{max} words). The reason for the insignificant improvement in CPU-time is that the DO-loop was already previously vectorized. There are then two cases to be considered: in the first case RECU is used with a vectorlength of (N-2) and in the other one - when RECU is used again in the same iteration after the boundary-layer thickness has been increased - with a vectorlength of one (NL=NU). If we assume an average value for N of about 80, the second case is far less time consuming than the first. In the new version of RECU, the vectorlength for block 1 is increased nine times to about 700 in the first case, and to 9 in the second. Now executing 3 statements nine times instead 27 statement once results in a relatively high improvement on the CRAY. Unfortunately, however, we do not achieve much of an improvement in the first case, where now 3 statements are executed 700 times instead of 27 statements 78 times. The latter would not be the case on a CYBER 205 where the increase in performance is especially large for very long vectors, but relatively poor for short vectors (see also chapter 2.3), whereas the

CRAY-1S has already very good performance on vectors shorter than , say 100.

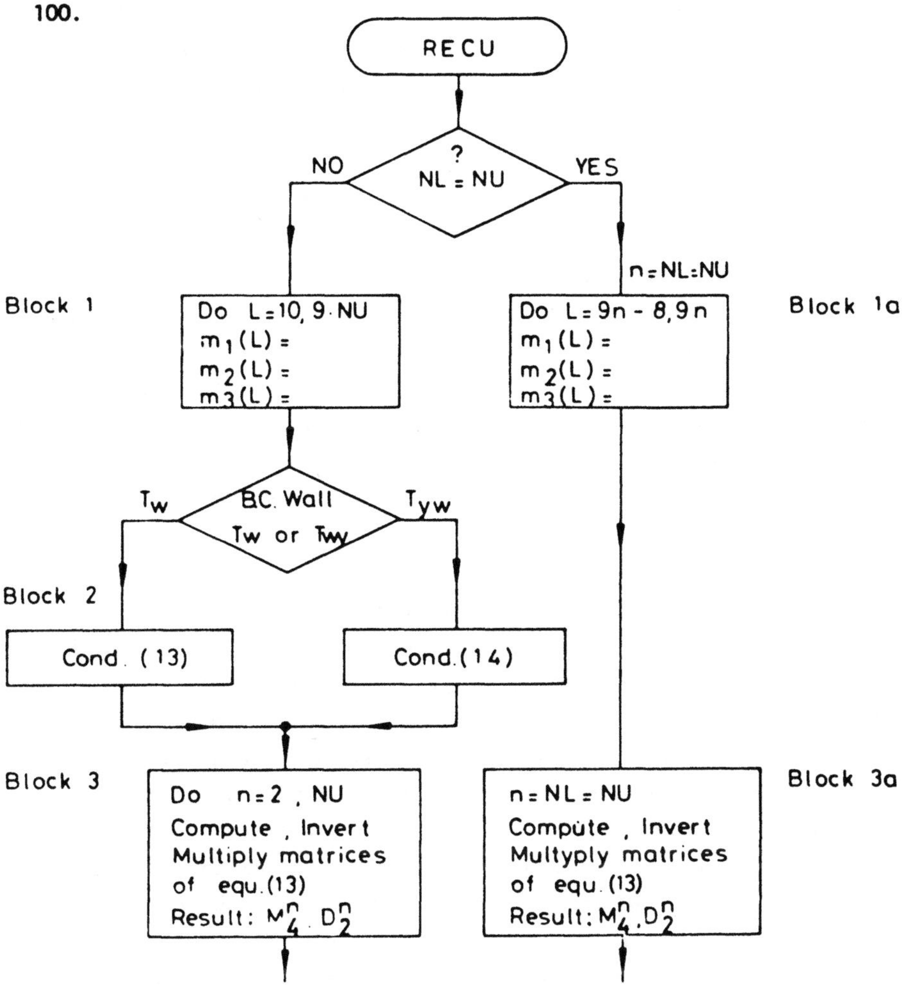

Fig. 3: Vectorized version of the subroutine RECU.

The next change in RECU performed easily, without increased storage requirements, but it too resulted in only a minimal improvement in performance (6. in Table 1). Figure 3 describes the new flow chart for RECU resulting from all the above changes. The last change is the splitting of the routine in two branches for the two cases mentioned above: one for NL = 2 and NU = N-1 and one for NL = NU when the boundary-layer thickness is increased, and where in the latter case, block 2 with the boundary conditions is not needed. Due to this splitting the DO-loop in block 1a is vectorized as a shortloop (i.e. for vectorlength less than 65), this being less time consuming than a normal vector loop. In block 3a a DO-loop is no longer needed, its start-up time then being saved.

The last change to be made is again in the routine FO3D We already mentioned that unformatted I/O is very fast on the CRAY, but we can reduce the CPU time even more using Buffer I/O instead of READ/WRITE I/O. This is due to the fact that Buffer I/O is performed in parallel to the computation whenever possible. The use of Buffer I/O is only recommended, however, if the data transfer in a single READ/WRITE statement is large (i.e. at least several hundred words). In the present case more then 28000 I/O operations with more than 1600 words each are performed with Buffer I/O instead of unformatted READ/WRITE, resulting in another 2.6 seconds saving in CPU-time (7. in Table 1).

Having dealt with different changes which can improve the performance of boundary-layer codes on a CRAY-1S, we find that the gain is about 25 per cent for routine RECU and about 50 per cent for routine FO3D, thus resulting in a reduction of about 40 per cent of the total computation time. Before comparing with the CPU time on the IBM 3081K we have to test the improved version of the code on this machine. Since the increased loop length in block 1 of RECU (5. in Table 1) is not believed to influence the time on a serial computer, the corresponding changes are not introduced in the IBM code. Also the buffer instructions were not used on the IBM. With this improved version of the code, the CPU time is 411 seconds i.e. the gain is less than 1 per cent versus the old version (8. and 9. in Table 1).

The improved code on the CRAY-1S is almost 14 times faster than the "improved" code on the IBM 3081K, the latter being barely significant. If we switch off the autovectorization facility of CRAY, the CPU time on the CRAY-1S is again 52 seconds, which is 8 times faster than on the IBM. This indicates that our improvements barely influence the serial performance (as the CPU times on the IBM have shown), and that vectorization gives an improvement of only 40 per cent. The latter results from the fact that boundary-layer codes are highly serial owing to the implicit formulation of the equations normal to the wall.

The only way to write a three-dimensional,boundary-layer code with high vectorization is to use a difference scheme using only data in one plane (e.g. x^1 = const.) where all data are known, to calculate the solution in the next plane (see |12|). There is one disadvantage then; such schemes are usually more restricted when taking stability into account, than for example the zig-zag scheme of Krause et al. |3| used in the present boundary-layer code.

References

1. Prandtl L.: Ueber Fluessigkeitsbewegung bei sehr kleiner Reibung. Verhandlg. III Intern. Math. Kongr. Heidelberg 1904, 484-491.
2. Krause E.: Numerical Solutions of the Boundary Layer Equations. AIAA Journal 5, 1967, pp. 1231-1237.
3. Krause E., Hirschel E.H., Bothmann Th.: Die numerische Integration der Bewegungsgleichungen dreidimensionaler laminarer, kompressible Grenzschichten.Fachtagung Aerodynamik,Berlin 1969,DGLR-Fachbuchreihe Bd. 3, Braunschweig, 1969.
4. Krause E., Hirschel E.H., Bothmann Th.: Differenzenformeln zur Berechnung dreidimensionaler Grenzschichten. DLR-FB-69-66, Sept. 1969.
5. Blottner F.G.: Finite difference methods for solution of boundary-layer equations. AIAA Journal, Vol.8, No. 2, Febr. 1970, 193-205.
6. Blottner F.G.: Computational techniques for boundary layers. AGARD Lecture Series 73, Febr. 1975.
7. Schwamborn D.:Laminare Grenzschichten in der Naehe der Anlegelinie an Fluegeln und fluegelaehnlichen Koerpern mit Anstellung. DFVLR-FB 81-31, Aug. 1981.
   Also: Laminar boundary layers in the vicinity of the attachment line on wings and winglike bodies at incidence. ESA-TT-752, Dec. 1982.
8. Robert K.: Higher-Order boundary-layer equations for three-dimensional, compressible flow. DLR-FB 77-36 (1977), 205-215, also also ESA-TT-518 (1979), 273-288.
9. Hirschel E.H., Kordulla W.: Shear flow in surface-oriented coordinates, in: Notes on Numerical Fluid Mechanics, Vol. 4,

Vieweg-Verlag Braunschweig, Wiesbaden, 1981.
10. Hockney R.W., Jesshope C.R.: Parallel computers, Adam Hilger, Bristol, 1981.
11. Sydow P.J.: Optimization guide.CRAY Computer System Technical Note SN-0220.
12. Kenichi Matsuno: A vector-oriented finite-difference scheme for calculating three-dimensional compressible laminar and turbulent boundary layers on practical wing configurations.AIAA Paper 81-1020 Proceedings of the AIAA CFO Conference, Palo Alto, Cal., June 22-23, 1981.

9.4 Vectorization of the Galerkin-Method

Roland Kessler, DFVLR

9.4.1 Introduction

For the numerical simulation of incompressible flows spectral methods were used in many studies. This class of numerical methods has particular advantages in solving instability and transition problems. Different forms of spectral methods and their applications in fluid mechanics were discribed in the book of Gottlieb and Orszag |1|. Recent developments in spectral methods make it possible to calculate flows in complex geometries as well as to simulate three-dimensional flows.Some examples of this applications were given in |2| and |3|.

A special type of spectral methods is the Galerkin method described in this contribution. This method enables us to solve the three-dimensional time-dependent Navier-Stokes equations in a rectangular region |4|. The method is fully implicit and exactly fulfills all boundary conditions. In chapter 9.4.3 the principles of the Galerkin method are formulated. The selection of the basis functions, the numerical solution of the nonlinear ordinary differential equations and the implementation of the algorithm is discussed. To exploit the high performance of a vector computer such as the CRAY-1S some modifications of the algorithm are necessary. The vectorization of two subroutines, which are consuming more than 98 per cent of the processing time yields a good performance of the Galerkin method on the CRAY-1S and is discussed in chapter 9.4.5.

To illustrate the numerical method, we examine the instabilities of a convective flow which is confined to a rectangular box heated from below. To simulate these instabilities, the three-dimensional,time-dependent Boussinesq equations were employed. The principle sketch of Figure 1a shows the convection box and the flow configuration of three convection rolls. The calculated streamline of the steady solution in Figure 1b shows the three-dimensional flow induced by the rigid side walls of the container. The main interest of the numerical simulation is to calculate the time dependent instabilities which occur in such a flow. The results of these calculations using the Galerkin method were given in |4|.

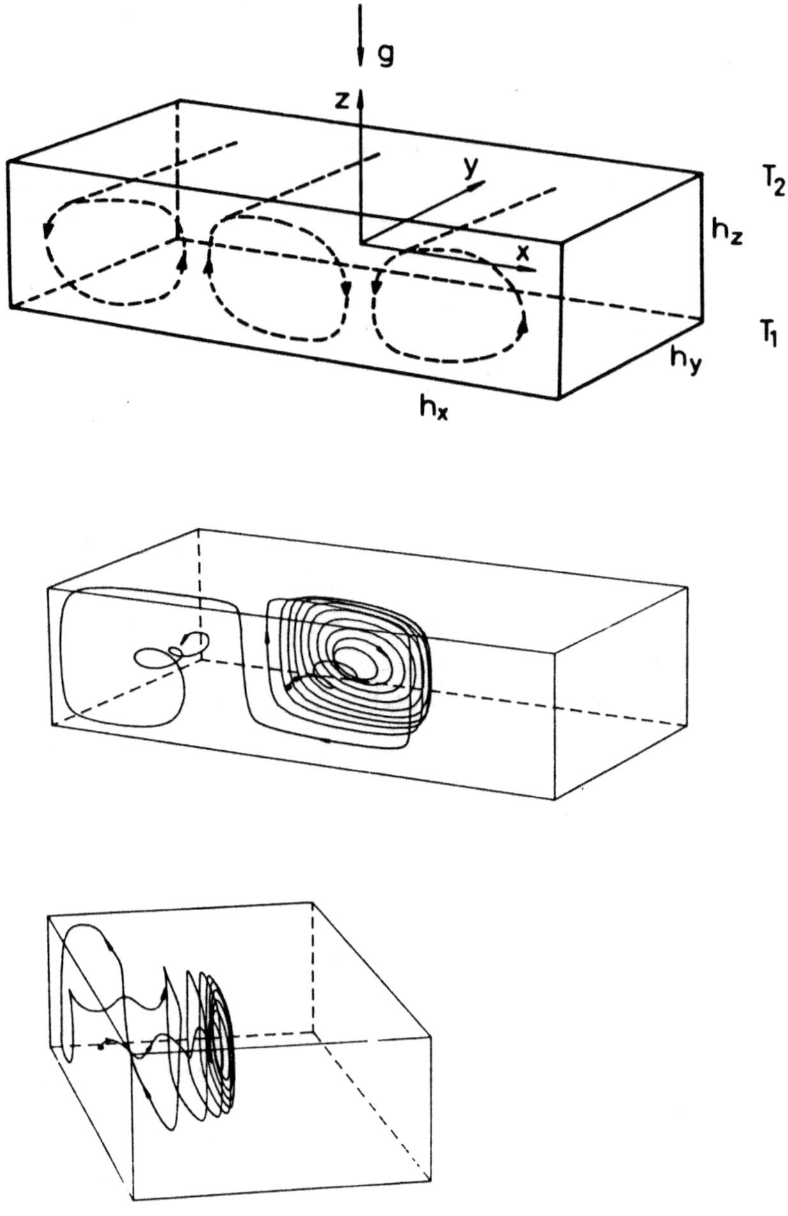

Fig. 1(a): Principle sketch and (b) calculated streamlines of steady convection flow in a box, $h_x:h_y:h_z$ = 4:2:1.

9.4.2 Basic Equations

Convective transport phenomena and their instabilities are described by the basic equations of continuum mechanics using the Boussinesq approximation. The fluid is considered incompressible except for the temperature dependence of the density in the buoyancy term causing the thermal convection. The remaining fluid properties are also assumed to be constant, the dissipation in the energy equation is neglected and the fluid is considered a Newtonian medium. The equations are nondimensionalized by introducing characteristic units: h_z the height of the fluid layer, h_z^2/κ the thermal diffusion time, $T_1 - T_2$ the temperature difference and $(\rho \nu \kappa)/h_z^3$ for length, time, temperature and the pressure. The conservation of mass, momentum and energy then leads to the following set of partial differential equations:

(1) $$\nabla \cdot \bar{v} = 0 ,$$

(2) $$\frac{1}{Pr}\left(\frac{\partial}{\partial t} + \bar{v} \cdot \nabla\right)\bar{v} = -\nabla p + \nabla^2 \bar{v} + Ra\, T\, \bar{e}_z ,$$

(3) $$\left(\frac{\partial}{\partial t} + \bar{v} \cdot \nabla\right) T = \nabla^2 T ,$$

$$Ra = \frac{\alpha g (T_1 - T_2) h_z^3}{\kappa \nu} , \qquad Pr = \frac{\nu}{\kappa} ,$$

with the velocity vector $\bar{v}=(u,v,w)$, pressure p, temperature T and the unit vector $\bar{e}_z=(0,0,1)$. The dimensionless parameters are the Prandtl number Pr and the Rayleigh number Ra.

We assume the no-slip condition at all walls of the rectangular box. The temperature at the upper and lower wall is assumed to be at the constant values T_2 and T_1. The side walls have been chosen either adiabatic or perfectly conducting.

9.4.3 Algorithm

In order to be able to solve the basic equations (1-3) with the Galerkin method we have to expand the unknow functions v,T in sets of given basis functions. These functions are continuous and defined in the whole computational region. The unknown coefficients of the basis functions were determined in such a way that the error is minimized in the computational region. The criterion for minimizing the error function gives the equations for the coefficients.

In the following the method will be briefly explained using a general formulation. A differential equation in the computational region V is defined with the operator L

(4) $$L(x,y,z,t,g,\dot{g},g'\ldots) = 0 .$$

The solution g of the differential equation is expanded into a complete set of linearly independent basis functions E_i with the time dependent coefficients $e_i(t)$

(5) $$\tilde{g} = \sum_{i=1}^{N} e_i(t)E_i(x,y,z) .$$

Each function E_i satisfies the boundary conditions of the differential equation L. Replacing g by the expansion (5) in the differential equation (4) we get the error function ε on the right hand side of the differential equation

(6) $$L(x,y,z,t,\sum_{i=1}^{N} e_i(t)E_i\ldots) = \varepsilon(x,y,z,t) .$$

To minimize the error function ε, we require the orthogonality of the error function with N linearly independent weighting functions w_i. The method of Galerkin is to employ the basis functions itself of the weighting functions w_i

(7) $$\int \varepsilon E_i \, dV = 0 \qquad i = 1,N .$$

The relation (7) leads to N equations for the unknown coefficients $e_i(t)$ called Galerkin equations, which in general form a system of ordinary differential equations.

To solve the Boussinesq equations (1-3) we expand the velocity vector v and the temperature T into sets of basis functions v_i and T_ℓ

(8)
$$\bar{v}(x,y,z,t) = \sum_{i=1}^{N} a_i(t)\bar{v}_i(x,y,z) ,$$
$$T(x,y,z,t) = \sum_{\ell=1}^{M} b_\ell(t)\, T_\ell(x,y,z) .$$

All the functions v_i and T_ℓ satisfy the boundary conditions exactly. In addition to this each basis function of the velocity vector v_i has zero divergency and so exactly fulfills the continuity equation. Therefore, the pressure term can be eliminated in the momentum equation. Applying the expansions (8) in the Boussinesq equations, the orthogonality relation (7) leads to the corresponding Galerkin equations, a system of nonlinear ordinary differential equations for the coefficient vectors a_i and b_l

(9)
$$A^{(1)}_{ki}\, a_i + Ra\, A^{(2)}_{k\ell}\, b_\ell - \frac{1}{Pr}\, A^{(3)}_{kij}\, a_i\, a_j = \frac{1}{Pr}\, A^{(4)}_{ki}\, \frac{\partial a_i}{\partial t} ,$$
$$B^{(1)}_{m\ell}\, b_\ell + B^{(2)}_{mi}\, a_i - B^{(3)}_{mi\ell}\, a_i\, b_\ell = B^{(4)}_{m\ell}\, \frac{\partial b_\ell}{\partial t} ,$$
$$i,j,k = 1,\dots,N \qquad \ell,m = 1,\dots,M ,$$

where A and B consist of integrals and are defined as follow:

(10)
$$A^{(1)}_{ki} = \int_V \bar{v}_k \cdot \Delta\bar{v}_\ell\, dV , \qquad B^{(1)}_{m\ell} = \int_V T_m \cdot \Delta T_\ell\, dV ,$$
$$A^{(2)}_{k\ell} = \int_V \bar{v}_k \cdot T_\ell \cdot \bar{k}\, dV , \qquad B^{(2)}_{mi} = \int_V T_m \cdot \bar{v}_i \cdot \bar{k}\, dV ,$$
$$A^{(3)}_{kij} = \int_V \bar{v}_k \cdot (v_i \cdot \nabla \cdot v_j)\, dV , \qquad B^{(3)}_{mi\ell} = \int_V T_m\, (\bar{v}_i \cdot \nabla \cdot T_\ell)\, dV ,$$
$$A^{(4)}_{ki} = \int_V \bar{v}_k \cdot \bar{v}_i\, dV , \qquad B^{(4)}_{m\ell} = \int_V T_m \cdot T_\ell\, dV .$$

These integrals depend only on the basis functions and are calculated numerically.

To solve the ordinary differential equations (9) the time derivatives were replaced by a finite difference operator. As the equation can be very stiff, explicit methods tend to numerical oscillations especially for large systems of equations. We therefore employ the so-called "One-leg-method", a derivation of the trapezoidal rule which was described by |5|:

$$\text{(11)} \qquad \frac{\partial u}{\partial t} = F(u) \longrightarrow \frac{u^{n+1} - u^{n}}{\Delta t} = F\left(\frac{u^{n+1} + u^{n}}{2}\right) .$$

This method is implicit, of second-order accuracy and remains stable up to large time steps. The Galerkin equations now take the following form:

$$A_{ki}^{(1)} a_i^{n+1/2} + Ra\, A_{k\ell}^{(2)} b_\ell^{n+1/2} - \frac{1}{Pr}\left(A_{kij}^{(3)} a_i^{n+1/2} a_j^{n+1/2} + A_{ki}^{(4)} (a_i^{n+1/2} - a_i^{n}) \frac{2}{\Delta t}\right) = 0 ,$$

$$\text{(12)} \quad B_{m\ell}^{(1)} b_\ell^{n+1/2} + B_{mi}^{(2)} a_i^{n+1/2} - B_{mi\ell}^{(3)} a_i^{n+1/2} b_\ell^{n+1/2} - B_{m\ell}^{(4)} (b_\ell^{n+1/2} - b_\ell^{n}) \frac{2}{\Delta t} = 0 ,$$

$$a_i^{n+1/2} = (a_i^{n+1} + a_i^{n})/2 , \quad b_\ell^{n+1/2} = (b_\ell^{n+1} + b_\ell^{n})/2 ,$$

where the arithmetically averaged values were replaced by $a_i^{n+\frac{1}{2}}$ and $b_\ell^{n+\frac{1}{2}}$. The Galerkin equations now were reduced to a system of nonlinear algebraic equations for the unknown coefficients a_i^{n+1}, b_ℓ^{n+1} of the new time level. These nonlinear equations were solved using a Newton method. To simplify the expressions we combine the coefficients a_i^n and b_ℓ^n to the vector $\bar{x}^n$ and replace the left hand side of equation (12) by the operator F_k:

$$\text{(13)} \qquad F_k(\bar{x}^{n+\frac{1}{2}}) = 0 .$$

By applying the Newton method we obtain the following iteration process for $\bar{x}^{n+1}$

$$\text{(14)} \qquad F_k(\bar{x}_k^{n+1/2}) + J(x_k^{n+1/2})(x_{k+1}^{n+1/2} - x_k^{n+1/2}) = 0$$

with the Jacobian matrix J

$$(15) \qquad J_{kj}(\bar{x}) = \frac{\partial F_k(\bar{x})}{\partial x_j} \; .$$

We now have to solve one linear system per iteration. The initial value $\bar{x}_0^{n+\frac{1}{2}}$ for the first iteration is determined by extrapolation of the values of $\bar{x}$ at the time steps n and n-1

$$(16) \qquad \bar{x}_0^{n+1/2} = \bar{x}^n + \frac{1}{2}\,(\bar{x}^n - \bar{x}^{n-1}) \; .$$

As the initial value is close to the solution of the nonlinear system (13) the accuracy is good enough after one iteration. The truncation error of the finite difference approximation in time is always larger than the error of the Newton method using only one iteration. So we have to solve one linear system per time step.

The structure of the Jacobian depends strongly on the basis functions chosen, and the values of the Rayleigh number and Prandtl number. As the matrix is full and the values of the diagonal are not dominant, an iterative method is not efficient to solve the linear system (14). We therefore use the Gaussian elimination.

The implementation of the Galerkin method is more complicated compared to corresponding finite difference schemes. Thus we confine ourselves to a discussion of the principle structure of the program. It can be divided into three main parts:

1. Calculation of the integrals.
2. Calculation of the Jacobian matrix.
3. Solving the linear system.

The number and the values of the integrals defined by (10) only depend on the basis functions. They remain constant during the whole calculation. Therefore there are two possibilities for the structure of the program.

The first version is to calculate the integrals at each time step simultaneously with the calculation of the Jacobian matrix. This allows us to realize a program with little storage requirement. Most of the storage is needed then for the Jacobian matrix. The total storage required is about $1.4*N^2$, were N is the number of basis functions. To simulate the three-dimensional, time-dependent convection flow 400 functions are necessary, corresponding to a relative little storage of 225000 words.

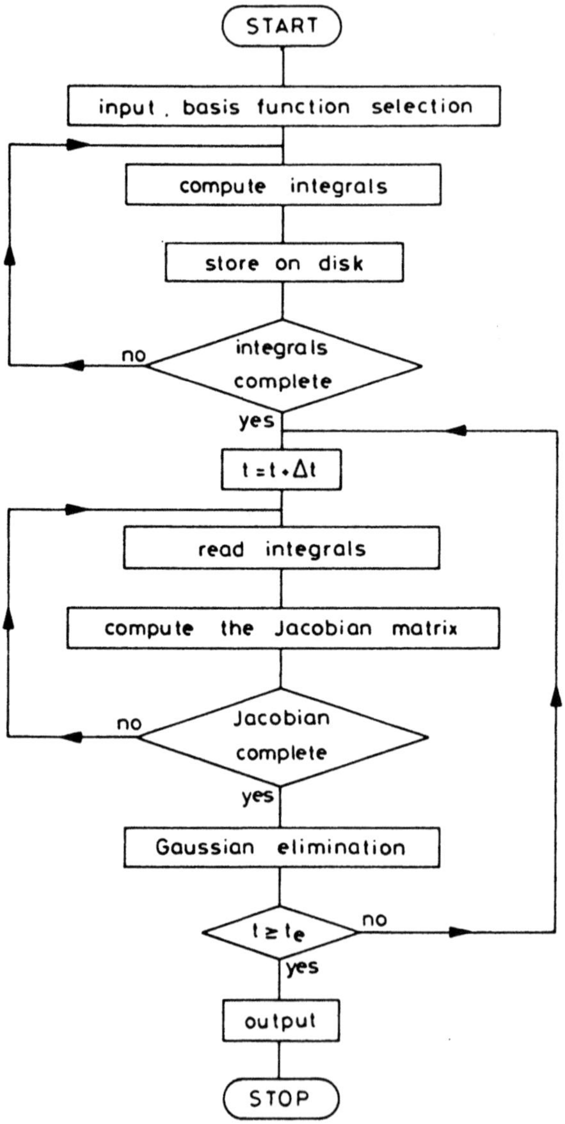

Fig. 2: Flow chart of the Galerkin code.

In the second version the integrals are calculated once and stored before the first time step. As the integrals are required at each time step, this saves a lot of computation time. According to the type of computer the computation time decreases by factors of about 2.5 to 7. In contrast to this advantage, the requirement of storage increases enormously. The number of integrals depends on the number of basis functions and the selection modus used. For 400 functions, about 7 to 8 million words have to be stored. As the memory of modern computers generally is not large enough, the values of the integrals have to be written on disk in blocks of about 270000 words. The data management is very easy, because every integral is required only once per time step.

We only realized the second version of the Galerkin method. The principle parts of the program are explained with the flow chart shown in Fig. 2. After the input of all parameters the integrals FINT were calculated in the same sequence as they were required later. After reaching the limit of 270000 the array FINT will be stored on disk and the next integrals can be calculated.

This operation will be repeated until all required integrals are calculated and stored on disk. Setting n equal to 1 the first time step begins. To determine the Jacobian matrix the values of the integrals are restored from disk to the array FINT again in blocks of 270000 words. After the Jacobian has been completely calculated the linear system is solved employing the Gaussian elimination, and the next time steps can be calculated until the final time t_e is reached.

9.4.4 Vectorization of the Galerkin method on the CRAY-1S

The code of the Galerkin method discribed above was written in FORTRAN IV and first implemented on an IBM 3081. We use both REAL * 4 and REAL * 8 to examine the influence of the round-off errors. The implementation on the CRAY-1S required only few modifications.

To vectorize the code efficiently, we employ the compiler option ON = F, which gives us information about the time required by each subroutine during the execution of the program. The result of this is summarized in Table 1.

Table 1: CRAY-1S CPU-time (in seconds) for the original version of the Galerkin code, 398 basis functions.

Input, Output, basic function selection	1.60	13.78 start-up time
Integrals, calculation and store	12.18	
Jacobian matrix	6.82	9.92 per time step
Subroutine DOVEK	5.47	
read integrals	1.30	
rest	0.05	
Gaussian elimination	3.10	

The input of parameters, the selection modus for the basis functions and all output utilities together require 1.60 seconds assuming a number of 398 basis functions. Before executing the first time step, we have to calculate all the integrals and to store them on disk. This part of the program requires 12.18 seconds. The code to calculate the Jacobian matrix is well structured and most of the time is spent in the subroutine DOVEK which contains only some nested DO-loops. Including the time for reading the integrals from disk, the total time required is 6.82 seconds. 3.10 seconds are needed for solving the linear system by Gaussian elimination.

In the sections of the code containing the selection of the basis functions and the calculation of the integrals, vectorization is very difficult. The selection modus for the basis functions is held very variable and there is no chance to get a good performance in these parts of the code. As we are interested mainly in time-dependent calculations with the need of 500 to 5000 time steps, the start-up time can be neglected compared to the time required for all time steps. Therefore we confine ourselves to the vectorization of the subroutine DOVEK and the Gaussian elimination.

As discussed above the integrals stored on disk have to be read in blocks of about 270000 words at each time step. In the example showed in Table 1 the total number of integrals is $7.3 * 10^6$. Using an implicit DO-loop in the READ-statement, which is vectorized on the CRAY-1S, the CPU requires 1.30 seconds to read the whole data. Replacing the implicit DO-loop by the BUFFER IN statement the time to read the data is reduced by a factor of 65 in this case. The 0.02 seconds now required to transfer the data can be neglected compared to the other operations per time step. Using this fast data transfer, we are no longer restricted by the relatively small memory of the CRAY-1S.

To solve the linear system defined by the Jacobian matrix we use a Gaussian elimination without pivoting.This is important for vectorization, because IF statements destroy the vector structure of the algorithm. Vectorization of Gaussian elimination has been discussed previously in this book (chapter 5.3). Employing the algorithm formulated there, we reduce the computational time by a factor of up to 3.8. Considering the calculation with 398 basis functions the elimination process now requires 0.81 seconds. Assuming $\frac{2}{3}N^3$ floating point operations to solve the N linear equations, we obtain a computing speed of 52 MFLOPS for N equal to 398.

As the Galerkin equations consist of sums of linear and quadratic terms only, the partial derivatives can be formulated directly to calculate the Jacobian matrix. The parts of the derivatives resulting from a linear term lead to a single loop, which is already vectorized. To calculate the parts resulting from the quadratic terms, two nested loops are required. These loops contain more than 98.5 per cent of the operations needed to build up the Jacobian matrix, and are integrated in the subroutine DOVEK. The whole task now is to vectorize this one subroutine.

In the original version the subroutine DOVEK contained the following nested loops

```
          DO  1  I = IA,IE
          DO  1  J = JA,JE
(17)      NB = NB + 1
          PA(I) = PA(I) + X(J) x FINT(NB)
        1 PA(J) = PA(J) + X(I) x FINT(NB)
```

The partial derivatives of one equation are stored in the array PA and the coefficients in the array X. To improve the performance of these loops we do the following steps:

a) Remove the dependencies
Vectorization of the inner loop yields wrong results if I is equal to J. Therefore the loop is not vectorized by the autovectorizer. To avoid the dependencies the inner loop is split into two loops. As the inner loops are now vectorized the performance improves by a factor of about 3.4.

b) Increase length of inner loops
As the basis functions of the velocity vector and the temperature are composed by 7 subsystems of different symmetry, the length of the inner loops varies from about 20 up to 100 even if all 400 functions are used. Defining a new array X1, we can copy two or three parts of the vector X to the new array to store the relevant parts continuously. Processing the longer vector X1, we can avoid especially the very shorts loops. Although we have to copy the vectors each time before processing the nested loops we can obtain a speed up factor of 1.3.

c) Use the CRAY-1S intrinsic functions
Considering the nested DO-loops in the example (17), we can recognize the different vector structures of the two statements in the inner loop. As the index I of the outer loop has a constant value for the inner loop, the linear expressions have the form

SCALAR = SCALAR + VECTOR * VECTOR
VECTOR = VECTOR + SCALAR * VECTOR

Therefore the expressions can be replaced by the intrinsic functions SDOT, which calculates the dot-product of two vectors and the function SAXPY which adds a vector and another scaled one. The improvement of performance by imploying these functions is not very high. Whereas the function SDOT has a speed-up factor of greater than one for all vector lengths, SAXPY is faster than the corresponding DO-loop only for a vector length greater than about 100.

Although nearly all operations to calculate the Jacobian matrix are done using intrinsic functions and therefore more than 99 per cent of this part of the code is vectorized, the performance is relatively poor. Employing the example with 398 basis functions, we can obtain a rate of only 30 MFLOPS. Considering the speed rates of the two intrinsic functions applied, this is well understood. The maximum performance of more than 60 MFLOPS is available only for a vector length of more than 300 for SAXPY and more than 1000 for SDOT. Employing loops with lengths in the range of 50 to 150 the function does not work very efficiently.

As there is no possibility to further increase the length of the vectors used by the algorithm, we have to change the DO-loops in an other manner. On the CRAY-1S the performance obtained strongly depends on the number of operations used in the inner loop. A modification which improves the performance of short loops considerably is described in the following section.

d) Unroll the outer loop partially

First we have to replace the intrinsic functions by the original inner DO-loops. To increase the number of operations in the inner loop, we are processing simultaneously the expressions for I, I+1, I+2 and I+3. Therefore the increment of the outer loop can be changed from 1 to 4. If the expression in the inner loop is a triadic operation of the form

VECTOR = VECTOR + SCALAR * VECTOR

this modification leads to rather high performance even for short vector lengths. To illustrate the modification, we show the original nested loops in comparison to the partially unrolled outer loop.

original

```
  DO   1   I = IA,IE
  DO   1   J = JA,JE
  NB = NB+1
1 PA(J) = PA(J) + X(I) * FINT(NB)
```

partially unrolled outer loop

```
    JD = JE+1-JA
    DO  1   I = IA,IE,4
    DO  2   J = JA,JE
    NB = NB+1
  2 PA(J) = (((PA(J) + X(I) * FINT(NB)) + X(I+1) * FINT(NB+JD))
  :       + X(I+2) * FINT(NB+2xJD)) + X(I+3) * FINT(NB+3xJD)
  1 NB = NB+3*JD
```

By using the parenthesis on the right hand side of the expression in the DO-loop we enhance chaining. If (IE-IA) is not a multiple of the increment 4, more operations are processed than in the original version. To avoid errors in this case the dimension of some arrays have to be increased and a copy of the vector X is necessary before processing the loop. Nevertheless the speed-up factor compared to the fully vectorized version with intrinsic functions is more than 1.7.

To unroll the outer loop is only profitable if the expression in the inner loop is of triadic form. Therefore we want to change all dot products to the triadic form. In our case this can easily be done by exchanging the inner and outer loop. This is illustrated by the following examples:

Dot product:

```
    DO  1   I = IA,IE
    DO  1   J = JA,JE
    NB = NB + 1
  1 PA(I) = PA(I) + X(J)*FINT(NB)
```

Triadic operation:

```
    JD = JE + 1 - JA
    DO  1   J = JA,JE
    NB = NB + 1
    II = 0
    DO  1   I = IA,IE
    II = II + 1
  1 PA(I) = PA(I) + X(J)*FINT(NB+JD*II)
```

Now we can partially unroll the outer loop again according to the previously discussed example. Although in this example the inner loop is now the short one, the performance of the partially unrolled loop is better than the version using the intrinsic function SDOT.

All nested loops in the subroutine DOVEK can now be formulated in the same manner with a partially unrolled outer loop. Employing the example with 398 basis functions, the performance improves to 51 MFLOPS for calculating the Jacobian matrix. Considering the asymptotic performance of 66 MFLOPS for the scalar product of very long vectors, this is a rather good result on the CRAY-1S.

9.4.5 Profit of Vectorization

The profit of all modifications described above will be shown by comparing the original version and the completely vectorized version on the CRAY-1S. In addition to this, the modified version is run with the autovectorizer switched off with the option OFF = V, that means no vectorization of the inner loops. The results of this comparison are shown in Table 2.

Table 2: Comparison of the original and vectorized versions, CRAY CPU-times in seconds for 398 basis functions and one time step.

	original version	vectorized version	vectorized version autovectorizer off
Input, Output basis function selection	1.60	1.60	2.37
Integrals, calculation and storage	12.18	10.75	12.62
Jacobian matrix	6.82	0.62	3.61
Gaussian elimination	3.10	0.81	6.69

As there are no modifications, the time required for input and output remains the same in the two versions. The decrease in time for calculating and storing the integrals is the profit of the BUFFER OUT statement. The time required for the calculations of the Jacobian matrix is decreased considerably by vectorization of the code. Also the Gaussian elimination used in the vectorized version has a speed-up factor of nearly four compared to the original version. By turning the autovectorizer off, we can see that only little is vectorized in the code computing the integrals and doing input and output. However, factors of 6 to 8 can be obtained by computing the Jacobian and the Gaussian elimination.

To get a feeling of the influence of the start-up time, we give a realistic example of solving the time dependent equations using 500 time steps. The CPU time and the relative importance of the sections are shown in Table 3.

Table 3: CPU-time on the CRAY-1S of the modified Galerkin code, 398 basis functions, 500 time steps (% : per cent).

	time (sec)	%
input, output	1.6	0.2
integrals	10.8	1.5
Jacobian matrix	309.6	42.5
Gaussian elimination	406.8	55.8
total	728.8	100.0

We can easily see that the start-up time required by input/output and the calculation of the integrals can be neglected and only the highly vectorized parts of the code are important.

In contrast to the CRAY-1S, which uses 64 bit words only, the IBM 3081 computer can use both 32 and 64 bit words. Performing the calculations in single precision accuracy on the IBM, round off errors of about 0.5 to 1 per cent are found in the results compared to the calculations using double precision. As the error produced by the Galerkin method is in the same range for three-dimensional calculations, we use single precision in the most cases on the IBM. Comparing the performance of the CRAY with the IBM 3081K, we therefore use single precision as well as double precision on the IBM.

Table 4: Speed-up factors of the Galerkin code, 398 basis functions, 500 time steps.

Cray 1S / IBM 3081 64 bit 32 bit	Cray 1S / IBM 3081 64 bit 64 bit
26.5	33.1

Assuming the example with 398 basis functions and 500 time steps, a speed up factor of 26.5 is found comparing the CRAY 64 bit version with the IBM 32 bit version. This factor increases up to 33.1 comparing the results using identical precision. The comparison makes obvious the high profit obtained by the vectorized Galerkin code on the CRAY-1S.

9.4.6 Conclusions

The vectorization of the time dependent Galerkin method has been illustrated by simulating three-dimensional convection flow in a rectangular box. In spite of the implicit nature of the method, the algorithm is vectorized very easily. As mentioned above, only two short subroutines have to be modified to obtain the high overall performance of more than 50 MFLOPS for the Galerkin code on the CRAY-1S. An essential condition to reach this high performance, is the very high transfer rate between the memory and the main storage using the buffered I/O. Employing this processor we are no longer restricted by the relatively small memory of the CRAY-1S. The vectorized Galerkin code illustrates that a high performance can be obtained even for a vector length 50 to 150 using many operations inside the inner DO-loops. Therefore the Galerkin code may obtain a higher performance on the CRAY-1S than on a CYBER 205 which requires long vector lengths.

References

1. Gottlieb D.O.,Orszag S.A.: Numerical analysis of spectral methods: theory and applications. NSF-CBMS Monograph No. 26, SIAM, Philadelphia 1977.
2. Orszag S.A.: Spectral methods for problems in complex geometries. J. Comp. Phys. 37, 1980, 70-92.
3. Kleiser L., Schumann U.: Spectral simulations of the laminar-turbulent transition process in plane poiseuille flow. Proc. Symposium on Spectral Methods for Partial Differential Equations. R.G. Voigt ed., SIAM, Philadelphia, Pa., 1983.
4. Kessler R.: Oszillatorische Konvektion. Doctoral thesis, Universitaet Karlsruhe, 1983.
5. Dahlquist G.: Error analysis for a class of methods of stiff non-linear initial value problems. Numerical Analysis Dundee 1975, Lecture Notes in Mathematics 506, A. Dold and B. Eichmann eds., Springer Verlag Berlin, 1976, 60-72.

9.5 Vectorization of the Direct Monte-Carlo Simulation

Eckart Meiburg, DFVLR

9.5.1 Physical Background

In fluid mechanics, as in many other disciplines the development of various regular structures has been observed and described over an extended time period. Its investigation contributes to the understanding of natural phenomena as well as further, technological development. For example, the wake of a body can have a certain structure, depending on its form and the parameters of the flow; this structure is the so-called Karman vortex street. It is composed of alternately left- and right-rotating vortices, which appear and flow downstream at a fixed frequency. Since vortices themselves are not visible, their existence can be observed only indirectly, such as by the humming of wires in the wind. Here the vortices separate from the wire, causing it to vibrate, this being eventually perceived as a sound.

Another striking example of the forces and possible consequences of such a regular flow structure is found in the collapse of the Tacoma Bridge near Seattle, Washington (USA). Separating vortices similarly brought the bridge to oscillations, which increased to the point of final collapse. The reason for the strong amplification of the oscillations is that the frequency of the separating vortices lay close to the resonant frequency of the bridge, thus enabling the wind to transfer a large amount of energy to the bridge.

Similar hazards are encountered in the construction of turbine and compressor blades, where resonance can be very destructive. The formation of structures in wakes of airplanes is also important; just as the vortices exerted forces upon the bridge, they can also act upon wings of aircraft, and so affect the manoeuvrability as well as the lift and drag coefficients, and thereby the efficiency. It is, therefore, an economic incentive to investigate the process leading to the formation of structures in wakes.

a)

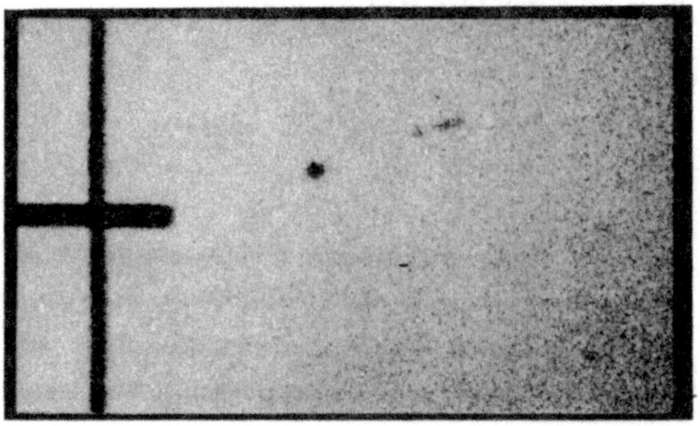

b)

c)

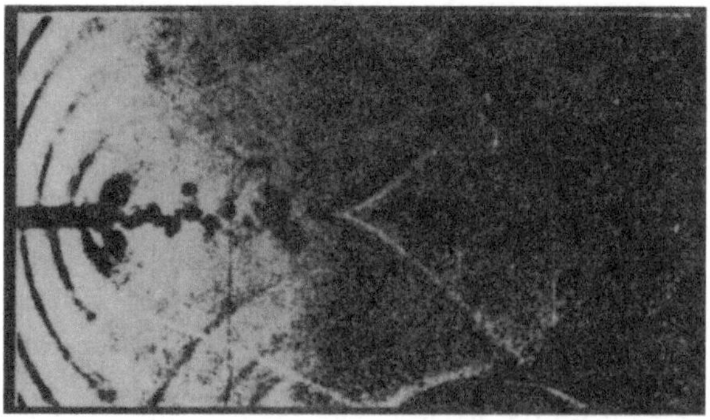

Fig. 1: Formation of vortices behind a flat plate |1| (Experiments of the German - French Institute ISL). a) Shock wave moving over the plate. b) Initial symmetrical vortex pair. c) Asymmetrical Karman vortex street and related pressure waves.

9.5.2 Configuration

The flow past a thick flat plate serves as a model for the theoretical experimental investigation of instabilities in wakes. Here the location of the separation is known and the simple geometry makes both the numerical treatment, and the experimental visualization of the flow possible. The extent to which a gaskinetic simulation can contribute to the understanding of the flow is investigated.

In both the experiment (Oertel |1|) and the numerical simulation, the flow is created by an incident shock (generated by a piston) which moves over the plate into the medium at rest. The no-slip condition at the plate causes boundary layers to form, which separate at the trailing edge, and so form vortices. First, a symmetrical vortex appears, which then becomes unstable and changes into the asymmetrical solution, the Karman vortex street (Fig. 1). The periodically separating vortices are related to pressure waves moving upstream.

9.5.3 The Governing Equation

The numerical simulation of the flow in the wake is based on the kinetic theory of gases as described by Hirschfelder, Curtis and Bird |2|, Kennard |3|, Ferziger and Kaper |4| and Chapman and Cowling |5|. This way we do not have to make any assumptions about the parameters of the medium, the coupling of the stress tensor with the velocity field, or about the modelling of turbulence. In contrast to continuum mechanics, the kinetic theory of gases describes the molecular structure of the gas by means of a distribution function $f(\vec{r},\vec{c},t)$. This tells us how the velocities $\vec{c}$ are distributed over the gas particles in space $\vec{r}$ and time t. Once f is known, the macroscopic flow quantities can be deduced. The distribution function is determined by the Boltzmann equation

$$\frac{\partial f}{\partial t} + \vec{c}\,\frac{\partial f}{\partial \vec{r}} + \vec{F}\,\frac{\partial f}{\partial \vec{c}} = \left(\frac{\partial f}{\partial t}\right)_{collisions} ,$$

which states that the rate of change of the distribution function in an element of the phase space $(\vec{r},\vec{c})$ is caused by convection of particles, external forces, and collisions among particles.

If we confine ourselves to binary collisions, the collision term on the right-hand side of the equation can be replaced by an integral, so that the Boltzmann equation becomes a non-linear, integro-differential equation. This must then be solved for the given initial and boundary conditions.

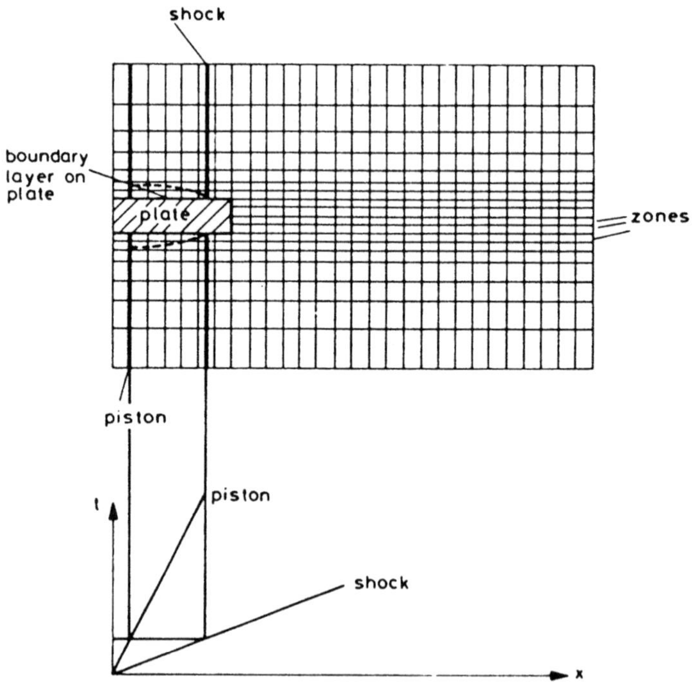

Fig. 2: Partitioning of the flowfield into zones and cells.

9.5.4 The Algorithm

The Boltzmann equation is solved numerically with the direct simulation Monte-Carlo method, as described by Bird |6|, and applied by him, Seiler |7|, and others. This method is based on the fact that the solutions to the Boltzmann equation can be made invariant by an appropriate normalization if the product of the number of particles, n, and their collision cross-section, δ, is constant (see Derzko |8|). This allows us to simulate the real gas

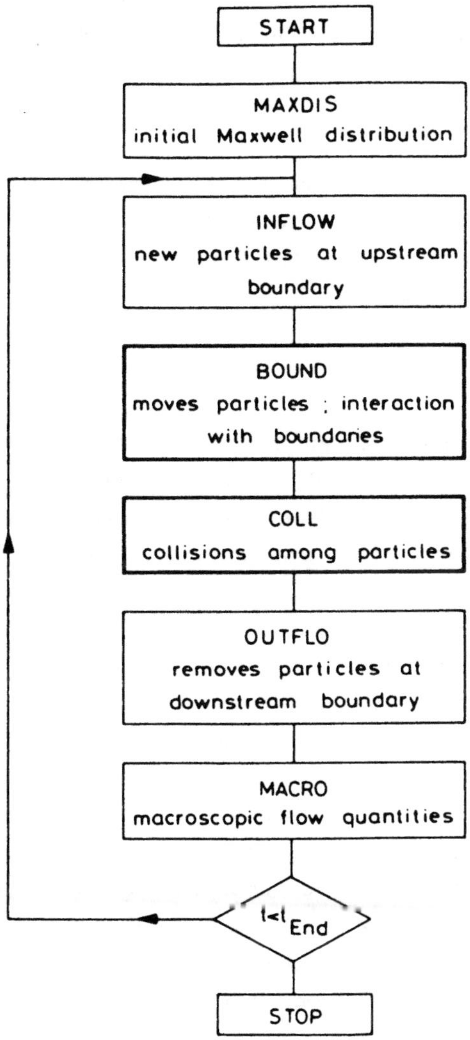

Fig. 3: Flow chart of the Monte-Carlo program.

flow with several thousand particles. For the simulation, the flowfield is divided into zones and cells (see Fig. 2), the size of which has to be such that the macroscopic flow quantities change little from one to the next. In the numerical simulation as in the shock-tube experiment, the flow around the plate is created by a piston which moves into the medium at rest from the left, and so produces a shock (see Fig. 2). So the process of the calculation is as follows (see Fig. 3). First in the subroutine MAXDIS the particles are positioned in the control volume with a Maxwell distribution of their velocities, so that the location and the velocity of each particle are determined by random variables. From now on the simulation proceeds in time steps Δt, each of which is composed of the following five parts:

1. Subroutine INFLOW

The piston which has moved a short distance into the medium, returns, at which point new particles are positioned in the empty space. Their number, mean velocity, and temperature are calculated from the shock equations. The location and velocity of each particle is again determined by random variables.

2. Subroutine BOUND

Here the new positions of the particles at the end of the time step are calculated from their old positions and velocities, such that their interaction with the boundaries (diffuse, specular or mixed reflection) is taken into account, but not the interaction among the particles.

3. Subroutine COLL

Here the collisions among the particles are computed. We confine ourselves to binary collisions, which dominate in low density gases. In each cell, two particles are selected randomly without taking into consideration their position within the cell. The probability of a collision between these particles is proportional to their relative velocity, so that not every pair that has been selected will collide. Every collision that has occurred corresponds to a time interval Δt_c, which depends on the relative velocity of the colliding particles and on the local particle density. Each cell has a time counter, which adds up the intervals Δt_c. A set of collisions appropriate to the time step Δt has occurred when the time counters of all cells have reached the level $m \cdot \Delta t_c$, where m is the number of the time step.

4. Subroutine OUTFLO

Here the constant density condition at the outflow boundary is satisfied by removing an appropriate number of particles from the flow.

5. Subroutine MACRO

It calculates the macroscopic flow quantities density, velocity, and temperature in each cell, by sampling the particle properties.

The Monte-Carlo method uncouples the motion of the particles and the interaction among them, so that the time step Δt has to be small. An increasing number of particles and a decreasing cell size also improve the results.

All variables are made dimensionless by the mean free path and the most probable velocity of the particles. The hard sphere model is applied for the calculation of the collisions.

9.5.5 Vectorization

It became obvious that in longer calculations more than 70 per cent of the CPU-time was used in the subroutine COLL and more than 99 per cent in COLL and BOUND, so that our efforts were directed towards an efficient vectorization of these subroutines.

The original structure of the subroutine COLL was strictly serial. It consisted of a DO-loop over all zones in which the following steps were carried out for each cell individually (see Fig. 4). First it had to be determined whether its time counter was already full, which could be the case if a very unlikely collision with a large Δt_c had occurred in the previous time step. If that was the case, then the next cell was treated; otherwise the number of particles in the cell was checked because it takes at least two to produce a collision. If this condition was also fulfilled, a collision pair for the cell was selected until a pair with an appropriate relative velocity was found or an upper bound for the number of selections was reached. The collision was then calculated for this pair, and then we determined with the time counter whether another collision should

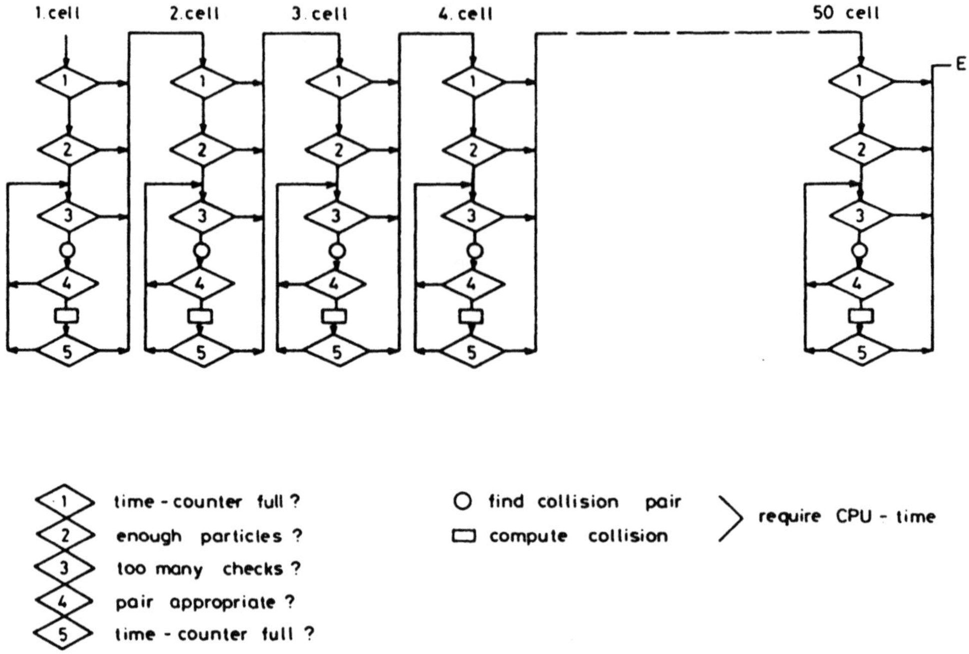

Fig. 4: Serial structure of the subroutine COLL.

occur in the cell or whether the calculation should proceed to the next cell. Thus the cells in a zone were all treated individually before the next zone was dealt with. It is obvious that the subroutine COLL contained a great deal of necessary IF-conditions related to GOTO-statements, which complicated the vectorization. But the selection of a collision pair, the calculation of its relative velocity and collision probability, as well as the collision of the particles, consist of numerous arithmetic operations, which can easily be vectorized. Therefore, the subroutine was restructured to that shown in Fig. 5. The goal was to form a vector of cells for the selection of the collision pair and the calculation of the collision. First the time counter and the number of particles in each cell are checked. The cells that fulfill these conditions are now formed into a vector. The random selection of the collision pair, as well as the calculation of the relative velocity and the collision probability can be carried out in vectorized form.

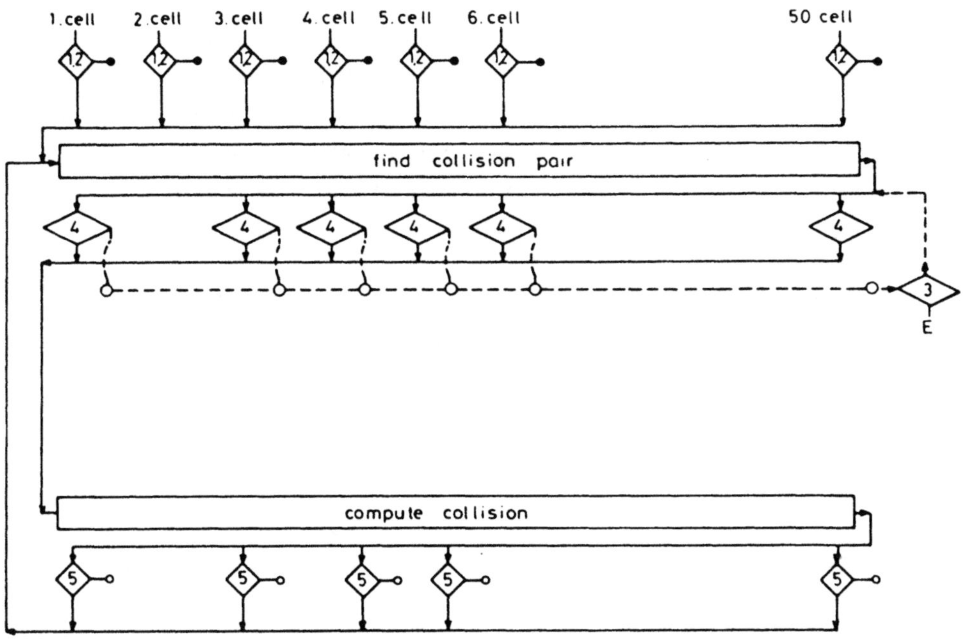

Fig. 5: Vectorized version of the subroutine COLL.

Another collision pair is selected for those cells not having one yet, so that a collision can be calculated in each cell of the vector. This selection of another collision pair is carried out in non-vectorized form, since it may take place in the cells 5,6,9, and 17, so that we do not have a constant increment. The effort of re-organising the data proved to be greater than the gain of the then possible vectorization.

Now a collision can be calculated in vectorized form for every cell in the vector. This means that the six post-collision velocity components must be determined from:

- the conservation of linear momentum in all three directions
- the conservation of angular momentum
- the conservation of energy
- the characteristics of the potential model chosen.

The calculation of the six post-collision velocity components proceeds as follows: first the deflection angle and the relative magnitude of two of the post-collision relative velocity components are determined by selecting two random numbers. Its third component now follows from the equation of conservation of energy. Finally the post-collision velocities of the particles are obtained by adding half the post-collision relative velocity to the mean velocity for one particle and subtracting it for the other one. These operations can be vectorized completely.

Now the time counter of each cell in the vector is checked. If it is already filled, another cell from the same zone is taken into the vector, for which the next collision will then be calculated. It is only when there are no more cells left untreated, that a collision can no longer be calculated for each cell in the vector. With 128 cells per zone, a vector length of 30 to 35 yielded the best results.

The subroutine BOUND can be vectorized only to a small degree. The operation

$$\vec{x}_{new} = \vec{x}_{old} + \vec{v} \cdot \Delta t$$

can be carried out in vectorized form, but the test of whether the particle collides with one of the boundaries or not is always related to a GOTO-statement, so that the CVMGT-routine cannot be applied.

Because the direct Monte-Carlo method is insensitive to the propagation of errors due to machine inaccuracy, the calculations on the IBM needed to be carried out only in single precision accuracy. In contrast to that, the lowest possible accuracy on the CRAY-1S corresponds to double precision accuracy on the IBM, so that the effective storage needed on the CRAY-1S is double of that of the IBM. This fact, along with the possibility of taking more particles for the simulation owing to the higher computational speed, requires that the particle positions and velocities be read from and written to mass storage at every time step. Here the BUFFER-statement proved to be very useful, since it can read and write the data while the calculation is in progress.

9.5.6 Comparison of CPU-times on the IBM 3081 and the CRAY-1S

Test calculations were carried out under three conditions:

1. single precision on the IBM
2. on the CRAY-1S without vectorization
3. on the CRAY-1S with vectorization.

The time step and the number of particles were also varied, so that their influence could be studied. CPU-times per time step are listed in Table 1.

Table 1: CPU-times (in seconds) for IBM 3081K and CRAY-1S.

particles per box	Δt	CPU [sec] IBM	Cray no vector.	Cray with vector.	rate of acceleration IBM/Cray no vector.	IBM/Cray with vector.
10	2	51.11	6.03	4.39	8.47	11.64
10	10	161.81	32.29	16.64	5.01	9.72
50	2	206.66	26.91	19.50	7.68	10.60

We see that for the calculation with 10 particles per box and a time step of 2 the fully vectorized CRAY version is 11.64 times faster than the IBM version. Most of this acceleration (8.47) is due to the high scalar performance of the CRAY-1S, whereas the vectorization contributes only relatively little. If we increase the time step, this ratio changes, with the scalar CRAY version only being about 5 times faster than the IBM version. This is due to the fact that different operations have different rates of scalar acceleration. The highly vectorized COLL-subroutine has now gained importance compared to BOUND, so that now the vectorization contributes almost a factor of 2 to the acceleration. If we increase the number of particles the number of particle-related operations increases proportionally. So the operations related to the zones, such as READ- and WRITE-statements, must be responsible for the fact that the CPU-time increases less than proportionally. On the IBM the INPUT and OUTPUT operations were done with implicit DO-loops, whereas on the CRAY-1S the very fast BUFFER IN and BUFFER OUT statements were used. The influence of the BUFFER and CVMGT commands were checked in separate calculations (Table 2).

Table 2: CPU-times (in seconds) for CRAY-1S for different versions of the code ((1): with BUFFER and CVMGT, (2): with WRITE and CVMGT, (3): with BUFFER and IF ; % : per cent).

particles per box	Δt	CPU-times [sec] (1)	(2)	(3)
10	2	4.39	4.69 (+6.8%)	4.61 (+5.0%)

Replacing BUFFER statements by WRITE statements slowed down the calculation by 6.8 per cent, and using IF statements instead of CVMGT statements required 5.0 per cent more CPU-time.

From the numbers given above we can draw the conclusion that with some programming effort the direct Monte-Carlo simulation can be accelerated by about an order of magnitude on the CRAY-1S. This rate could be improved if it were possible to perform single precision calculations and if more memory were available. The BUFFER and CVMGT-commands proved to be valuable tools in vectorizing the program.

References

1. Oertel H.: Vortices in Wakes induced by shockwaves. Proc. Shock Tube Conference 1983.
2. Hirschfelder J.O., Curtis C.F., Bird R.B.: Molecular theory of gases and liquids. New York (1954).
3. Kennard E.H.: Kinetic theory of gases. London (1938).
4. Ferziger J.H., Kaper H.G.: Mathematical theory of transport processes in gases. Amsterdam-London (1972).
5. Chapman S., Cowling T.G.: The mathematical theory of non-uniform gases. Cambridge (1960).
6. Bird G.A.: Molecular Gas Dynamics. Oxford University Press (1976).
7. Seiler F.: Stossstruktur in Wandnaehe, Ph. D. Thesis, Karlsruhe (1980).
8. Derzko N.A.: Review of Monte-Carlo Methods in kinetic theory. UTIAS Review 35 (1972).

GPSR Compliance

The European Union's (EU) General Product Safety Regulation (GPSR) is a set of rules that requires consumer products to be safe and our obligations to ensure this.

If you have any concerns about our products, you can contact us on ProductSafety@springernature.com

In case Publisher is established outside the EU, the EU authorized representative is:

Springer Nature Customer Service Center GmbH
Europaplatz 3
69115 Heidelberg, Germany

Batch number: 10683525

Printed by Printforce, the Netherlands